CAMBRIDGE LIBRARY COLLECTION

Books of enduring scholarly value

Earth Sciences

In the nineteenth century, geology emerged as a distinct academic discipline. It pointed the way towards the theory of evolution, as scientists including Gideon Mantell, Adam Sedgwick, Charles Lyell and Roderick Murchison began to use the evidence of minerals, rock formations and fossils to demonstrate that the earth was older by millions of years than the conventional, Bible-based wisdom had supposed. They argued convincingly that the climate, flora and fauna of the distant past could be deduced from geological evidence. Volcanic activity, the formation of mountains, and the action of glaciers and rivers, tides and ocean currents also became better understood. This series includes landmark publications by pioneers of the modern earth sciences, who advanced the scientific understanding of our planet and the processes by which it is constantly re-shaped.

Illustrations of the Huttonian Theory of the Earth

John Playfair (1748–1819) was a Scottish mathematician and geologist best known for his defence of James Hutton's geological theories. He attended the University of St Andrews, completing his theological studies in 1770. In 1785 he was appointed joint Professor of Mathematics at the University of Edinburgh, and in 1805 he was elected Professor of Natural Philosophy. This highly influential book, first published in 1802, contains Playfair's clarification and summary of Hutton's geological concepts. Playfair concisely explains Hutton's theories on erosion and geothermal heat in rock formation and the concept of uniformitarianism in geology, illustrating these theories with his own precise observations on different types of rock strata. The clarity of Playfair's explanations was instrumental in popularising Hutton's geological theories, many of which are now recognised as key principles of modern geology. Playfair's strident defence of Hutton's ideas formed part of a controversial debate between Hutton's supporters and his detractors.

Cambridge University Press has long been a pioneer in the reissuing of out-of-print titles from its own backlist, producing digital reprints of books that are still sought after by scholars and students but could not be reprinted economically using traditional technology. The Cambridge Library Collection extends this activity to a wider range of books which are still of importance to researchers and professionals, either for the source material they contain, or as landmarks in the history of their academic discipline.

Drawing from the world-renowned collections in the Cambridge University Library, and guided by the advice of experts in each subject area, Cambridge University Press is using state-of-the-art scanning machines in its own Printing House to capture the content of each book selected for inclusion. The files are processed to give a consistently clear, crisp image, and the books finished to the high quality standard for which the Press is recognised around the world. The latest print-on-demand technology ensures that the books will remain available indefinitely, and that orders for single or multiple copies can quickly be supplied.

The Cambridge Library Collection will bring back to life books of enduring scholarly value (including out-of-copyright works originally issued by other publishers) across a wide range of disciplines in the humanities and social sciences and in science and technology.

Illustrations of the Huttonian Theory of the Earth

JOHN PLAYFAIR

CAMBRIDGE UNIVERSITY PRESS

Cambridge, New York, Melbourne, Madrid, Cape Town,
Singapore, São Paolo, Delhi, Tokyo, Mexico City

Published in the United States of America by Cambridge University Press, New York

www.cambridge.org
Information on this title: www.cambridge.org/9781108072311

This edition first published 1802
This digitally printed version 2011

ISBN 978-1-108-07231-1 Paperback

ILLUSTRATIONS

OF THE

HUTTONIAN THEORY

OF THE EARTH.

BY JOHN PLAYFAIR,

F. R. S. EDIN. AND PROFESSOR OF MATHEMATICS

IN THE UNIVERSITY OF EDINBURGH.

Nunc naturalem causam quærimus et assiduam, non raram et fortuitam.

SENECA.

EDINBURGH:

PRINTED FOR CADELL AND DAVIES, LONDON, AND

WILLIAM CREECH, EDINBURGH.

1802.

ADVERTISEMENT.

THE Treatiſe here offered to the Public, was drawn up with a view of explaining Dr Hutton's Theory of the Earth in a manner more popular and perſpicuous than is done in his own writings. The obſcurity of theſe has been often complained of; and thence, no doubt, it has ariſen, that ſo little attention has been paid to the ingenious and original ſpeculations which they contain.

THE ſimpleſt way of accompliſhing the object propoſed, ſeemed to be, to preſent a General Outline of the Syſtem, in one continued Diſcourſe; and to introduce afterwards, in the form of Notes, what farther Elucidation any particular ſubject was thought to demand. Through the whole, I have aimed at little more than a clear expoſition of facts, and a plain deduction of the concluſions grounded on them; nor ſhall I claim any merit to myſelf, if, in the order which I have found it neceſſary to adopt, ſome arguments may have taken a new form, and ſome additions may have been made to a ſyſtem naturally rich in the number and variety of its illuſtrations.

Of the qualifications which this undertaking requires, there is one that I may ſafely ſuppoſe myſelf to poſſeſs. Having been inſtructed by Dr Hutton himſelf in his theory of the earth; having lived in intimate friendſhip with that excellent man for ſeveral years, and almoſt in the daily habit of diſcuſſing the queſtions here treated of; I have had the beſt opportunity of underſtanding his views, and becoming acquainted with his peculiarities, whether of expreſſion or of thought. In the other qualifications neceſſary for the illuſtration of a ſyſtem ſo extenſive and various, I am abundantly ſenſible of my deficiency, and ſhall therefore, with great deference, and conſiderable anxiety, wait that deciſion from which there is no appeal.

Edinburgh College,
1ſt March 1802.

TABLE

OF

CONTENTS.

2. Whinſtone.

SECTION III.

PHENOMENA COMMON TO STRATIFIED AND UN-STRATIFIED BODIES. Page 97

NOTES

NOTES AND ADDITIONS.

ſhire

2. Granite

NOTE

IL-

ERRATA.

Page 44. *line* 4. *from the bottom*, *for* that *read* as
—— 189. —— 6. —————— *for* appearanes. *read* appearances.
—— 464. —— 4. —————— *for* D'AUBENTON *read* DAUBENTON
—— 482. —— 12. —————— *for* adversaries *read* adversary

ILLUSTRATIONS, &c.

A VERY little attention to the phenomena of the mineral kingdom, is ſufficient to convince us, that the condition of the earth's ſurface has not been the ſame at all times that it is at the preſent moment. When we obſerve the impreſſions of plants in the heart of the hardeſt rocks; when we diſcover trees converted into flint, and entire beds of limeſtone or of marble compoſed of ſhells and corals; we ſee the ſame individual in two ſtates, the moſt widely different from one another; and, in the latter inſtance, have a clear proof, that the preſent land was once deep immerſed under the waters of the ocean. If to this we add, that many maſſes of rock, the moſt ſolid and compact, conſiſt of no other materials but ſand and gravel; that, on the other hand, looſe gravel, ſuch as is formed only in beds of rivers, or on the ſea-ſhore, now abounds in places remote from both: if we reflect, at the ſame time, on the irregular

A and

and broken figure of our continents, and the identity of the mineral ſtrata on oppoſite ſides of the ſame valley, or the ſame inlet of the ſea; we ſhall ſee abundant reaſon to conclude, that the earth has been the theatre of many great revolutions, and that nothing on its ſurface has been exempted from their effects.

To trace the ſeries of theſe revolutions, to explain their cauſes, and thus to connect together all the indications of change that are found in the mineral kingdom, is the proper object of a THEORY OF THE EARTH.

But, though the attention of men may be turned to the theory of the earth by a very ſuperficial acquaintance with the phenomena of geology, the formation of ſuch a theory requires an accurate and extenſive examination of thoſe phenomena, and is inconſiſtent with any but a very advanced ſtate of the phyſical ſciences. There is, perhaps, in thoſe ſciences, no reſearch more arduous than this; none certainly where the ſubject is ſo complex; where the appearances are ſo extremely diverſified, or ſo widely ſcattered, and where the cauſes that have operated are ſo remote from the ſphere of ordinary obſervation. Hence the attempts to form a theory of the earth are of very modern origin, and as, from the ſimplicity of its ſubject, aſtronomy is the eldeſt, ſo, on account of the complexneſs

plexneſs of its ſubject, geology is the youngeſt of the ſciences.

It is foreign from the preſent purpoſe, to enter on any hiſtory of the ſyſtems that, ſince the riſe of this branch of ſcience, have been invented to explain the phenomena of the mineral kingdom. It is ſufficient to remark, that theſe ſyſtems are uſually reduced to two claſſes, according as they refer the origin of terreſtrial bodies to FIRE or to WATER; and that, conformably to this diviſion, their followers have of late been diſtinguiſhed by the fanciful names of *Vulcaniſts* and *Neptuniſts*. To the former of theſe Dr HUTTON belongs much more than to the latter; though, as he employs the agency both of fire and of water in his ſyſtem, he cannot, in ſtrict propriety, be arranged with either.

In the ſuccinct account which I am now about to give of this ſyſtem, I ſhall conſider the mineral kingdom as divided into two parts, namely, ſtratified and unſtratified ſubſtances. I ſhall treat, firſt, of the phenomena peculiar to the ſtratified; next, of thoſe peculiar to the unſtratified; and, laſtly, of the phenomena common to both. Beginning, then, with the firſt, the ſubject naturally divides itſelf into three branches; viz. the *materials*, the *conſolidation*, and the *poſition* of the ſtrata.

SECTION I.

OF THE PHENOMENA PECULIAR TO STRATIFIED BODIES.

1. *Materials of the Strata.*

1. IT is well known that, on removing the loose earth which forms the immediate surface of the land, we come to the solid rock, of which a great proportion is found to be regularly disposed in strata, or beds of determinate thickness, inclined at different angles to the horizon, but separated from one another by equidistant superficies, that often maintain their parallelism to a great extent. These strata bear such evident marks of being deposited by water, that they are universally acknowledged to have had their origin at the bottom of the sea; and it is also admitted, that the materials which they consist of, were then either soft, or in such a state of comminution and separation, as rendered them capable of arrangement by the action of the water in which they were immersed. Thus far most of the theories of the earth agree; but

but from this point they begin to diverge, and each to affume a character and direction peculiar to itfelf. Dr Hutton's does fo, by laying down this fundamental propofition, That in all the ftrata we difcover proofs of the materials having exifted as elements of bodies, which muft have been deftroyed before the formation of thofe of which thefe materials now actually make a part *.

2. The calcareous ftrata are the portion of the mineral kingdom that gives the cleareft teftimony to the truth of this affertion. They often contain fhells, corals, and other exuviæ of marine animals in fo great abundance, that they appear to be compofed of no other materials. Though thefe remains of organized bodies are now converted into ftone or into fpar, their fhape and interior ftructure are often fo well preferved, that the fpecies of animal or plant of which they once made a part, can ftill be diftinguifhed and pointed out among the living inhabitants of the ocean.

Others of the calcareous ftrata appear to be compofed of fragments of fome ancient rocks, which, after having been broken, have been again united into a compact ftone. In thefe we find pieces clearly marked as having been once continuous, but now placed at a diftance from

* Hutton's Theory, vol. i. p. 20. &c.

one another, and exhibiting exactly the ſame appearances as if they floated in a fluid of the ſame ſpecific gravity with themſelves.

From theſe, therefore, and a variety of ſimilar appearances, Dr Hutton concludes, that the materials of all the calcareous ſtrata have been furniſhed, either from the diſſolution of former ſtrata, or from the remains of organized bodies. But, though this concluſion is meant to be extended to all the calcareous ſtrata, it is not aſſerted that every cubic inch of marble or of limeſtone contains in it the characters of its former condition, and of the changes through which it has paſſed. It may, however, be ſafely affirmed, that there is ſcarce any entire ſtratum where ſuch characters are not to be found. Theſe muſt be held as deciſive with reſpect to the whole ſyſtem of ſtrata to which they belong; they prove the exiſtence of calcareous rocks before the formation of the preſent; and, as the deſtruction of thoſe is evidently adequate to the ſupply of the materials of theſe that we now ſee, to look for any other ſupply were ſuperfluous, and could only embarraſs our reaſonings by the introduction of unneceſſary hypotheſes *.

3. The ſame concluſions reſult from an examination of the ſiliceous ſtrata; under which we may comprehend the common ſand-ſtone, and

* Note I.

and alſo thoſe pudding-ſtones or breccias where the gravel conſiſts of quartz. In all theſe inſtances, it is plain, that the ſand or gravel exiſted in a ſtate quite looſe and unconnected, at the bottom of the ſea, previous to its conſolidation into ſtone. But ſuch bodies of gravel or ſand could only be formed from the attrition of large maſſes of quartz, or from the diſſolution of ſuch ſand-ſtone ſtrata as exiſt at preſent; for it will hardly be alleged, that ſand is a cryſtallization of quartz, formed from that ſubſtance, when it paſſes from a fluid to a ſolid ſtate.

Thoſe pudding ſtones in which the gravel is round and poliſhed, carry the concluſion ſtill farther, as ſuch gravel can only be formed in the beds of rivers or on the ſhores of the ſea; for, in the depths of the ocean, though currents are known to exiſt, yet there can be no motion of the water ſufficiently rapid to produce the attrition required to give a round figure and ſmooth ſurface to hard and irregular pieces of ſtone. There muſt have exiſted, therefore, not only a ſea, but continents, previouſly to the formation of the preſent ſtrata.

The ſame thing is clearly ſhewn by thoſe petrifactions of wood, where, though the vegetable ſtructure is perfectly preſerved, the whole maſs is ſiliceous, and has, perhaps, been found in the heart of ſome mountain, deep imbedded in the ſolid rock.

4. Characters of the ſame import are alſo found among the argillaceous ſtrata, though perhaps more rarely than among the calcareous or ſiliceous. Such are the impreſſions of the leaves and ſtems of vegetables; alſo the bodies of fiſh and amphibious animals, found very often in the different kinds of argillaceous ſchiſtus, and in moſt inſtances having the figure accurately preſerved, but the ſubſtance of the animal replaced by clay or pyrites. Theſe are all remains of ancient ſeas or continents; the latter of which have long ſince diſappeared from the ſurface of the earth, but have ſtill their memory preſerved in thoſe archives, where nature has recorded the revolutions of the globe.

5. Among bituminous bodies, pit-coal is the only one which conſtitutes regular and extenſive ſtrata; and no foſſil has its origin from the waſte of former continents, marked by ſtronger and more diſtinct characters. Not to mention that the coal ſtrata are alternated with thoſe that have been already enumerated, and that they often contain ſhells and corals, perfectly mineralized, it is ſufficient to remark, that there are entire beds of this foſſil, which appear to conſiſt wholly of wood, and in which the fibrous ſtructure is perfectly preſerved. From theſe inſtances,

ſtances, the appearances of vegetable ſtructure may be traced through all poſſible gradations, down to an evaneſcent ſtate. This laſt ſtate is undoubtedly the moſt common; and though coal does not then, on bare inſpection, make known its vegetable origin, yet, if we take it in connection with the other terms of the ſeries, as we may call them; if we conſider that the two extremes, viz. coal, with the vegetable ſtructure perfect, and coal without any ſuch ſtructure viſible, are often found in the ſame or in contiguous beds; and, if we remark, that through all theſe gradations coal contains nearly the ſame chemical elements, and yields, on analyſis, bitumen and charcoal, combined with a greater or leſs proportion of earth: if we take all theſe circumſtances into account, we cannot doubt that this foſſil is every where the ſame, and derives its origin from the trees and plants that grew on the ſurface of the earth before the formation of the preſent land.

6. Dr Hutton has further obſerved, that if thoſe ancient continents were at all ſimilar to the preſent, we can be at no loſs to account for the want of any diſtinct mark of vegetable organization in the greater part of the coal ſtrata. It is plain, that the daily waſte of animal and vegetable ſubſtances on the ſurface of the earth, muſt diſengage a great quantity of oily as well

as

as carbonic matter, which, with whatever element it is at firft combined, is ultimately delivered into the ocean. Thus, the oily or fuliginous parts of animal and vegetable fubftances, let loofe by burning, firft afcend into the atmofphere, but are at length precipitated, and either fall immediately into the fea, or are, in part at leaft, wafhed down into it from the land. From other caufes alfo, much vegetable matter is carried down by the rivers; and the whole quantity of animal and vegetable fubftances thus delivered into the fea, muft be very confiderable, amounting annually to the whole refiduum of thofe fubftances, not employed in the maintenance or reproduction of animal and vegetable bodies. Whether chemically united to the waters of the ocean, or fimply fufpended in them, this matter is at laft precipitated, and, mingling with earthy fubftances, is formed into ftrata, the place of which will be determined by the currents, the pofition of the prefent continents, and many other circumftances not eafily enumerated.

If, then, an order of things fimilar to what we now fee, exifted before the formation of the prefent ftrata, it would neceffarily happen, that the animal and vegetable fubftances, diffufed through the ocean, being feparated from the water, would be depofited at the bottom of the fea, ſea, and, in the courſe of ages, would form beds, leſs or more pure, according to the quantity of earth and other ſubſtances depoſited at the ſame time. Theſe beds being conſolidated and mineralized by operations that are afterwards to be conſidered, have been converted into pit-coal, the parts of which are impalpable, and retain nothing of their primitive ſtructure *.

If, then, the formation of coal from animal and vegetable bodies be admitted, the general poſition which derives the origin of the ſtrata from the waſte of former land, as it is applicable to all the kinds already enumerated, and of courſe to all thoſe with which they are alternated, comprehends a very large portion of the earth's ſurface. It comprehends, indeed, all the ſtrata uſually diſtinguiſhed by the name of *Secondary*; but there is another great diviſion of the mineral kingdom, viz. the rocks, called *Primitive*, which, as they are never alternated with the ſecondary, but are always inferior to them, muſt be further examined, before we can decide whether the ſame concluſion extends to them or not.

7. Here it muſt be carefully obſerved, that, among the primary rocks, the granite is not meant to be included, except where that ſtone is ſtratified, and either coincides with veined granite

* Note ii.

granite or with gneiſs. The primitive ſtrata, in Dr Hutton's theory, comprehend, beſides gneiſs, the micaceous, chlorite, hornblend, and ſiliceous ſchiſtus, together with ſlate, and ſome other kinds of argillite ; to which we muſt add, ſerpentine, micaceous limeſtone, and the greater part of marbles. Theſe are moſtly diſtinguiſhed by their laminated ſtructure, by having their planes much elevated with reſpect to the horizon, and by belonging more to the mountainous than the level parts of the earth's ſurface. They rarely contain veſtiges of organized bodies ; ſo rarely, indeed, that they were called primitive by the geologiſts who firſt diſtinguiſhed them from other rocks, on the ſuppoſition of their being part of the primeval nucleus of the globe, which had never undergone any change whatſoever ; but this, I believe, has now almoſt ceaſed to be the opinion of any geologiſt *. The Neptuniſts hold the rocks, here enumerated, and alſo granite, to be produced by aqueous depoſition ; but maintain them to be in the ſtricteſt ſenſe primeval, and of a formation antecedent to all organized bodies.

8. In oppoſition to this, Dr Hutton maintained, that the primary ſchiſtus, like all the other ſtrata, was formed of materials depoſited at the bottom

* Note III.

bottom of the ſea, and collected from the waſte of rocks ſtill more ancient. When, therefore, he conformed to the received language of mineralogiſts, by calling theſe ſtrata primitive, he only meant to deſcribe them as more ancient than any other ſtrata now exiſting, but not as more ancient than any that ever had exiſted. They are diſtinguiſhed, in his ſyſtem, by the name of *Primary*, rather than of *Primitive* ſtrata.

That the account now given of their origin is well founded, may be proved by unqueſtionable facts. For, firſt, though, agreeably to the obſervation juſt made, the ancient ſtrata do but rarely contain any remains of organized bodies, they are not entirely deſtitute of them. Different places in this iſland have been pointed out by Dr Hutton, where marine objects have been diſcovered in primary limeſtone, either by himſelf or others, and it would not be difficult to add more inſtances of the ſame kind*. In Dauphiny, coal, which is certainly a derivative ſubſtance, has been found among mountains which have a title to the character of primitive, ſuch as no one will diſpute. Theſe facts put the compoſition of ſuch rocks from looſe materials, beyond all doubt, and alſo prove their formation to be poſterior to the exiſtence of

* Note iv.

of an animal and vegetable ſyſtem. They do indeed prove this in the ſtricteſt ſenſe, only of the particular beds in which they are found; but as theſe beds are in all other reſpects as much to be accounted primary as any part of the mineral kingdom, it is evident that the negative inſtances are here of no force, and that nothing can be gained to the adverſaries of this opinion by denying it in general, if they are obliged to admit it in a ſingle caſe.

9. Again, it is certain, as Dr Hutton remarks, that there are few conſiderable bodies of ſchiſtus, even the moſt decidedly primitive, where ſand and gravel may not in ſome parts be obſerved. Indeed, it is not only true that they are to be found in ſome parts of them; but, in fact, among many of the primitive mountains, we find large tracts, compoſed entirely of a ſchiſtoſe and much indurated ſand-ſtone, in beds highly inclined, ſometimes alone, ſometimes alternated with other ſchiſti. In many of them, the ſand of which they conſiſt appears to be entirely of granite, from the detritus of which rock it ſhould ſeem that they were chiefly formed.

10. Thus we conclude, that the ſtrata both primary and ſecondary, both thoſe of ancient and thoſe of more recent origin, have had their materials furniſhed from the ruins of former continents, from the diſſolution of rocks, or the

deſtruction

deſtruction of animal or vegetable bodies, ſimilar, at leaſt in ſome reſpects, to thoſe that now occupy the ſurface of the earth. This concluſion is not indeed proved of every individual portion of rock, but it is demonſtrated of many and large parts, and thoſe ſcattered indifferently through all the varieties of the ſtrata; and therefore, from the rules of the ſtricteſt reaſoning, we muſt infer, that the whole is derived from the ſame origin *.

Thus far concerning the materials of the ſtrata; and, as theſe were originally looſe and unconnected, we muſt next conſider by what means they were conſolidated into ſtone.

2. *Conſolidation of the Strata.*

11. Though Dr Hutton has no where defined the meaning of the term conſolidation, he has been ſcrupulouſly exact in uſing it conſtantly in the ſame ſenſe. He underſtands by it, not merely that quality in a hard body by which its parts cohere together, but alſo that by which it fills up the ſpace comprehended within its ſurface, being to ſenſe without poroſity, and impervious to air and moiſture.

Now,

* NOTE V.

Now, a porous maſs of unconnected materials, ſuch as the ſtrata appear originally to have been, can acquire hardneſs and ſolidity only in two ways, that is, either when it is firſt reduced by heat into a ſtate of fuſion, or at leaſt of ſoftneſs, and afterwards permitted to cool; or when matter that is diſſolved in ſome fluid menſtruum, is introduced along with that menſtruum into the porous maſs, and, being depoſited, forms a cement by which the whole is rendered firm and compact. Fire and water, therefore, are the only two phyſical agents to which we can aſcribe the conſolidation of the ſtrata; and, in order to determine to which of them that effect is to be attributed, we muſt inquire whether there are any certain characters that diſtinguiſh the action of the one from that of the other, and which may be compared with the phenomena actually obſerved among mineral ſubſtances.

12. Firſt, then, it is evident, that the conſolidation produced by the action of water, or of any other fluid menſtruum, in the manner juſt referred to, muſt neceſſarily be imperfect, and can never entirely baniſh the poroſity of the maſs. For the bulk of the ſolvent, and of the matter it contained in ſolution, being greater than the bulk of either taken ſingly, when the latter was depoſited, the former would have ſufficient room left, and would continue to oc-

cupy

cupy a certain ſpace in the interior of the ſtrata. A liquid ſolvent therefore could never ſhut up the pores of a body to the entire excluſion of itſelf; and, had mineral ſubſtances been conſolidated, as here ſuppoſed, the ſolvent ought either to remain within them in a liquid ſtate, or, if evaporated, ſhould have left the pores empty, and the body pervious to water. Neither of theſe, however, is the fact; many ſtratified bodies are perfectly impervious to water, and few mineral ſubſtances contain water in a liquid ſtate. That they ſometimes contain it, chemically united to them, is no proof of their ſolidity having been brought about by that fluid; for ſuch chemical union is as conſiſtent with the ſuppoſition of igneous as of aqueous conſolidation, ſince the region in which the fire was applied, on every hypotheſis, muſt have abounded with humidity.

13. Again, if water was the ſolvent by which the conſolidating matter was introduced into the interſtices of the ſtrata, that matter could conſiſt only of ſuch ſubſtances as are ſoluble in water, whereas it conſiſts of a vaſt variety of ſubſtances, altogether inſoluble either in it, or in any ſingle menſtruum whatſoever. The ſtrata are conſolidated, for example, by quartz, by fluor, by feltſpar, and by all the metals, in their endleſs

 combinations

combinations with ſulphureous bodies. To affirm that water was ever capable of diſſolving theſe ſubſtances, is to aſcribe to it powers which it confeſſedly has not at preſent; and, therefore, it is to introduce an hypotheſis, not merely gratuitous, but one which, phyſically ſpeaking, is abſurd and impoſſible.

This is not all, however; for, even if this difficulty were to be paſſed over, it would ſtill be required to explain, how the water, which, together with the matter which it held in ſolution, had inſinuated itſelf into the pores of the ſtrata, became ſuddenly diſpoſed to depoſite that matter, and to allow it, by cryſtallization or concretion, to aſſume a ſolid form *. The Neptuniſts muſt either aſſign a ſufficient reaſon for this great and univerſal change, or muſt expect to ſee their ſyſtem treated as an inartificial accumulation of hypotheſes which aſſigns oppoſite virtues to the ſame ſubject, and is alike at variance with nature and with itſelf; in a word, a ſyſtem that might paſs for the invention of an age, when as yet ſound philoſophy had not alighted on the earth, nor taught man that he is but the miniſter and interpreter of nature, and can neither extend his power nor his knowledge

* Note VI.

ledge a hair's-breadth beyond his experience and obſervation of the preſent order of things *.

14. Such are the more obvious, but I think unanſwerable objections, that may be urged againſt the aqueous conſolidation of the ſtrata. It is true, that ſtony concretions, ſome of them much indurated, are formed in the humid way under our eyes. Very particular conditions, however, are required for that purpoſe, and conditions ſuch as can hardly have exiſted at the bottom of the ſea. Firſt, The water muſt diſſolve the ſubſtance of which the concretion is to be formed, as it actually does in the caſe of calcareous, and in certain circumſtances, in that of ſiliceous, earth. Secondly, It muſt be ſeparated from that ſubſtance, as by evaporation, or by a combination of the matter diſſolved with ſome third ſubſtance, to which it has a greater affinity than to water, ſo as to form with it an inſoluble compound. Laſtly, The water that is deprived of its ſolution muſt be carried off, and more of that which contains the ſolution muſt be ſupplied, as ſometimes happens

 where

* Homo naturæ miniſter, et interpres tantùm facit et intelligit, quantùm de naturæ ordine re, vel mente, obſervaverit: nec amplius ſcit, aut poteſt.

Nov. Org. lib. i. aph. 1.

where water runs in a ſtream, or drops from the roof of a cavern. The two laſt conditions are peculiarly inapplicable to the bottom of the ſea, where the ſtate of the ſurrounding fluid would neither permit the water that was deprived of its ſolution from being drawn off, nor that which contained the ſolution from ſucceeding it.

It is further to be obſerved, that the conſolidation of ſtalactitical concretions, that is, the filling up of their pores, is always imperfect, and is brought about by the repeated action of the fluid running through the porous maſs, and continuing to depoſite there ſome of the matter it holds in ſolution. This, which is properly infiltration, is incompatible with the nature of a fluid, either nearly, or altogether quieſcent.

15 In order to judge whether objections of equal weight can be oppoſed to the hypotheſis of igneous conſolidation, we muſt attend to a very important remark, firſt made by Dr Hutton, and applied with wonderful ſucceſs to explain the moſt myſterious phenomena of the mineral kingdom.

It is certain, that the effects of fire on bodies vary with the circumſtances under which it is applied to them, and therefore a conſiderable allowance muſt be made, if we would compare

pare the operation of that element when it consolidated the ſtrata, with the reſults of our daily experience. The materials of the ſtrata were diſpoſed, as we have already ſeen, looſe and unconnected, at the bottom of the ſea; that is, even on the moſt moderate eſtimation, at the depth of ſeveral miles under its ſurface. At this depth, and under the preſſure of a column of water of ſo great a height, the action of heat would differ much from that which we obſerve here upon the ſurface; and, though our experience does not enable us to compute with accuracy the amount of this difference, it nevertheleſs points out the direction in which it muſt lie, and even marks certain limits to which it would probably extend.

The tendency of an increaſed preſſure on the bodies to which heat is applied, is to reſtrain the volatility of thoſe parts which otherwiſe would make their eſcape, and to force them to endure a more intenſe action of heat. At a certain depth under the ſurface of the ſea, the power even of a very intenſe heat might therefore be unable to drive off the oily or bituminous parts from the inflammable matter there depoſited, ſo that, when the heat was withdrawn, theſe principles might be found ſtill united to the earthy and carbonic parts, forming a ſubſtance very unlike the reſiduum obtained after combuſtion under a preſſure no greater than the weight of the atmoſphere. It is in like manner reaſonable to believe, that, on the application of heat to calcareous bodies under great compreſſion, the carbonic gas would be forced to remain; the generation of quicklime would be prevented, and the whole might be ſoftened, or even completely melted; which laſt effect, though not directly deducible from any experiment yet made, is rendered very probable, from the analogy of certain chemical phenomena.

16. An analogy of this kind, derived from a property of the barytic earth, was ſuggeſted by that excellent chemiſt and philoſopher, the late Dr Black. The barytic earth, as is well known, has a ſtronger attraction for fixed air than common calcareous earth has, ſo that the carbonate of barytes is able to endure a great degree of heat before its fixed air is expelled. Accordingly, when expoſed to an increaſing heat, at a certain temperature, it is brought into fuſion, the fixed air ſtill remaining united to it: if the heat be further increaſed, the air is driven off, the earth loſes its fluidity, and appears in a cauſtic ſtate. Here, it is plain, that the barytic earth, which is infuſible, or very refractory, *per ſe*, as well as the calcareous, owes its fuſibility to the preſence of the fixed air; and it is therefore

fore probable, that the same thing would happen to the calcareous earth, if by any means the fixed air were prevented from escaping when great heat is applied to it. This escape of the fixed air is exactly what the compression in the subterraneous regions is calculated to prevent, and therefore we are not to wonder if, among the calcareous strata, we find marks of actual fusion having taken place *.

17. These effects of pressure to resist the decomposition, and augment the fusibility of bodies, being once supposed, we shall find little difficulty in conceiving the consolidation of the strata by heat, since the intervals between the loose materials of which they originally consisted may have been closed, either by the softening of those materials, or by the introduction of foreign matter among them, in the state of a fluid, or of an elastic vapour. No objection to this hypothesis can arise from the considerations stated in the preceding case; the solvent here employed would want no pores to lodge in after its work was completed, nor would it find any difficulty in making its retreat through the densest and most solid substances in the mineral kingdom. Neither can its incapacity to dissolve the bodies submitted to its action be alleged. Heat is the most powerful and most general of all solvents; and, though some

* NOTE VII.

ſome bodies, ſuch as the calcareous, are able to reſiſt its force on the ſurface of the earth, yet, as has juſt been ſhewn, it is perfectly agreeable to analogy to ſuppoſe, that, under great preſſure, the carbonic ſtate of the lime being preſerved, the pureſt limeſtone or marble might be ſoftened, or even melted. With reſpect to other ſubſtances, leſs doubt of their fuſibility is entertained; and though, in our experiments, the refractory nature of ſiliceous earth has not been completely ſubdued, a degree of ſoftneſs and an incipient fuſion have nevertheleſs been induced.

Thus it appears, in general, that the ſame difficulties do not preſs againſt the two theories of aqueous and of igneous conſolidation; and, that the latter employs an agent incomparably more powerful than the former, of more general activity, and, what is of infinite importance in a philoſophical theory, vaſtly more definite in the laws of its operation.

18. A more particular examination of the different kinds of foſſils will confirm this concluſion, and will ſhow, that, wherever they bear marks of having been fluid, theſe marks are ſuch as characterize the fluidity of fuſion, and diſtinguiſh it from that which is produced by ſolution in a menſtruum. Dr Hutton has enumerated many of theſe diſcovered in the courſe of

that

that careful and accurate examination of foſſils, in which he probably never was excelled by any mineralogiſt. It will be ſufficient here to point out a few of the moſt remarkable examples.

19. Foſſil-wood, penetrated by ſiliceous matter, is a ſubſtance well known to mineralogiſts; it is found in great abundance in various ſituations, and frequently in the heart of great bodies of rock. On examination, the ſiliceous matter is often obſerved to have penetrated the wood very unequally, ſo that the vegetable ſtructure remains in ſome places entire; and in other places is loſt in a homogeneous maſs of agate or jaſper. Where this happens, it may be remarked, that the line which ſeparates theſe two parts is quite ſharp and diſtinct, altogether different from what muſt have taken place, had the flinty matter been introduced into the body of the wood, by any fluid in which it was diſſolved, as it would then have pervaded the whole, if not uniformly, yet with a regular gradation. In thoſe ſpecimens of foſſil-wood that are partly penetrated by agate, and partly not penetrated at all, the ſame ſharpneſs of termination may be remarked, and is an appearance highly characteriſtic of the fluidity produced by fuſion.

20. The round nodules of flint that are found in chalk, quite inſulated and ſeparate from

from one another, afford an argument of the ſame kind; ſince the flinty matter, if it had been carried into the chalk by any ſolvent, muſt have been depoſited with a certain degree of uniformity, and would not now appear collected into ſeparate maſſes, without any trace of its exiſtence in the intermediate parts. On the other hand, if we conceive the melted flint to have been forcibly injected among the chalk, and to have penetrated it, ſomewhat as mercury may, by preſſure, be made to penetrate through the pores of wood, it might, on cooling, exhibit the ſame appearances that the chalk-beds of England do actually preſent us with.

The ſiliceous pudding-ſtone is an inſtance cloſely connected with the two laſt; in it we find both the pebbles, and the cement which unites them, conſiſting of flint equally hard and conſolidated; and this circumſtance, for which it is impoſſible to account by infiltration, or the inſinuation of an aqueous ſolvent, is perfectly conſiſtent with the ſuppoſition, that a ſtream of melted flint has been forcibly injected among a maſs of looſe gravel.

21. The common grit, or ſandſtone, though it certainly gives no indication of having poſſeſſed fluidity, is ſtrongly expreſſive of the effects of heat. It is ſo, eſpecially in thoſe inſtances where the particles of quartzy ſand, of which

which it is composed, are firmly and closely united, without the help of any cementing substance whatsoever. This appearance, which is very common, seems to be quite inconsistent with every idea of consolidation, except an incipient fusion, which, with the assistance of a suitable compression, has enabled the particles of quartz to unite into stone.

It has indeed been asserted, that the mere apposition of stony particles, so as to permit their corpuscular attraction to take place, was sufficient to form them into stone. To this Dr Hutton has very well replied, that, admitting the possibility of a hard and firm body being produced in this way, of which, however, we have no proof, the close and compact texture, the perfect consolidation of the stones we are now speaking of, would still remain to be explained, and of this it is evident that the mere apposition of particles, and the force of their mutual attraction, can afford no solution.

22. These proofs that the strata must have endured the action of intense heat, though immediately deduced from those of the siliceous genus only, extend in reality to all the strata, of every kind, with which they are found alternated. It is impossible that heat, of the intensity here supposed, can have acted on a particular

lar ſtratum, and not on thoſe that are contiguous to it; and, as there are no ſtrata of any kind with which the quartzy and ſiliceous are not intermixed, ſo there are none of which the igneous conſolidation is not thus rendered probable. We need reſt nothing, however, on this argument, as the foſſils of every genus may be ſhewn to ſpeak diſtinctly for themſelves.

23. Thoſe of the calcareous genus do ſo perhaps more ſparingly than the reſt; yet even among them there are many facts, that, though taken unconnected with all others, are ſufficient to eſtabliſh the action of ſubterraneous fire. Such, for example, are the calcareous breccias, compoſed of fragments of marble or limeſtone, and not only adapted to each other's ſhape, but indented into one another, in a manner not a little reſembling the *ſutures* of the human *cranium*. From ſuch inſtances, it is impoſſible not to infer the ſoftneſs of the calcareous fragments when they were conſolidated into one maſs. Now, this ſoftneſs could be induced only by heat; for it muſt be acknowledged, that the action of any other ſolvent is quite inadequate to the ſoftening of large fragments of ſtone, without diſſolving them altogether.

24. In many other inſtances it appears certain, that the ſtones of the calcareous genus have been reduced by heat into a ſtate of fluidity

dity much more perfeƈt. Thus, the ſaline or finer kinds of marble, and many others that have a ſtruƈture highly cryſtallized, muſt have been ſoftened to a degree little ſhort of fuſion, before this cryſtallization could take place. Even the petrifaƈtions which abound ſo much in limeſtones, tend to eſtabliſh the ſame faƈt; for they poſſeſs a ſparry ſtruƈture, and muſt have acquired that ſtruƈture in their tranſition from a fluid to a ſolid ſtate *.

25. In accounting, by the operation of heat, for theſe appearances of fluidity, Dr Hutton has proceeded on the principle already laid down, as conformable to analogy, that calcareous earth, under great compreſſion, may have its fixed air retained in it, notwithſtanding the aƈtion of intenſe heat, and may, by that means, be reduced into fuſion, or into a ſtate approaching to it. In all this, I do not think that he has departed from the ſtriƈteſt rules of philoſophical inveſtigation. The faƈts juſt ſtated prove, that limeſtone was once ſoft, its fragments retaining at the ſame time their peculiar form, an effeƈt to which we know of none ſimilar but thoſe of fire; and therefore, though we could not conjeƈture how heat might be applied to limeſtone ſo as to melt it, inſtead of reducing it to a calx, we ſhould, neverthele ſs, have been forced to ſuppoſe

* Note viii.

ſuppoſe, that this had actually taken place in the bowels of the earth; and was a fact which, though we were not able to explain it, we were not entitled to deny. The principle juſt mentioned relieves us therefore from a difficulty, that would have embarraſſed, but could not have overturned, this theory of the earth.

26. From the arguments which the argillaceous ſtrata afford for the igneous conſolidation of foſſils, I ſhall ſelect one on which Dr Hutton uſed to lay conſiderable ſtreſs, and which ſome of the adverſaries of his ſyſtem have endeavoured to refute. This argument is founded on the ſtructure of certain iron-ſtones called *ſeptaria*, often met with among the argillaceous ſchiſtus, particularly in the vicinity of coal. Theſe ſtones are uſually of a lenticular or ſpheroidal form, and are divided in their interior into diſtinct *ſepta*, by veins of calcareous ſpar, of which one ſet are circular and concentric, the other rectilineal; diverging from the centre of the former, and diminiſhing in ſize as they recede from it. Now, what is chiefly to be remarked is, that theſe veins terminate before they reach the ſurface of the ſtone; ſo that the matter with which they are filled cannot have been introduced from without by infiltration, or in any other way whatſoever. The only other ſuppoſition, therefore, that is left for explaining the ſingular ſtructure

of

of this fossil, is, that the whole mass was originally fluid, and that, in cooling, the calcareous part separated from the rest, and afterwards crystallized.

27. It has been urged against this theory of the septaria, that these stones are sometimes found with the calcareous veins extending all the way to the circumference, and of course communicating with the outside. But it must be observed, that this fact does not affect the argument drawn from specimens in which no such communication takes place. It is at best only an ambiguous instance, that may be explained by two opposite theories, and may be reconciled either to the notion of igneous or of aqueous consolidation: but if there is a single close septarium in nature, it can, of course, be explained only by one of these theories, and the other must, of necessity, be rejected. Besides, it is plain, that a close septarium can never have been open, though an open septarium may very well have been close; and indeed, as this stone is, in certain circumstances, subject to perpetual exfoliation, it would be wonderful if no one was ever found with the calcareous veins reaching to the surface. With regard to the light, therefore, that they give into their own history, these two kinds of septaria are by no means on an equal footing; and this may serve to shew, how necessary

neceſſary it is, in all inductive reaſoning, and particularly in a ſubject ſo complex as geology, to ſeparate with care ſuch phenomena as admit of two ſolutions, from ſuch as admit only of one.

28. The bituminous ſtrata come next to be conſidered; and they are of great conſequence in the preſent argument, becauſe their diſſimilarity in ſo many particulars to all other mineral ſubſtances, renders them what Lord BACON calls an *inſtantia ſingularis*, having the firſt rank among facts ſubſervient to inductive inveſtigation. But though unlike in ſubſtance to other foſſils, and compoſed, as has been ſhewn, of materials that belonged not originally to the mineral kingdom, they agree in many material circumſtances with the ſtrata already enumerated. Their beds are diſpoſed in the ſame manner, and are alternated indiſcriminately with thoſe of all the ſecondary rocks, and, being formed in the ſame region, muſt have been ſubject to the ſame accidents, and have endured the operation of the ſame cauſes. They are traverſed too like the other ſtrata, by veins of all the metals, of ſpar, of baſaltes, and of other ſubſtances; and, whatever argument may hereafter be derived from this to prove the action of fire on the ſtrata ſo traverſed, is as much applicable to coal as to any other mineral. The coal ſtrata

ta alſo contain pyrites in great abundance, a ſubſtance that is perhaps, more than any other, the decided progeny of fire. This compound of metal and ſulphur, which is found in mineral bodies of every kind, I believe, without any exception, is deſtroyed by the contact of moiſture, and reſolved into a vitriolic ſalt. At the ſame time it is found in the ſtrata, not traverſing them in veins, which may be ſuppoſed of more recent formation than the ſtrata themſelves; but exiſting in the heart of the moſt ſolid rocks, often nicely cryſtallized, and completely incloſed, on all ſides, without the moſt minute vacuity. The pyrites muſt have been preſent, therefore, when the ſtrata were conſolidated, and it is inconceivable, if their conſolidation was brought about in the wet way, that a ſubſtance ſhould be ſo generally found in them, the very exiſtence of which is incompatible with humidity. This argument for the igneous origin of the ſtrata is applicable to them all, but eſpecially to thoſe of coal, as abounding with pyrites more than any other.

29. The difficulty that here naturally preſents itſelf, viz. how vegetable matter, ſuch as coal is ſuppoſed to have been, could be expoſed to the action of intenſe heat, without being deprived of its inflammable part, is obviated by the principle formerly explained concerning the effects

 of

of compreſſion. The weight incumbent on the ſtrata of coal, when they were expoſed to the intenſe heat of the mineral regions, may have been ſuch as to retain the oily and bituminous, as well as ſulphureous parts, though the whole was reduced almoſt to fuſion ; and thus, on cooling, the ſulphur uniting with iron might cryſtallize, and aſſume the form of pyrites.

30. The compreſſion, however, has not in every inſtance preſerved the bituminous, in union with the carbonic part of coal ; and hence a mark of the operation of fire quite peculiar to this foſſil, and found in thoſe infuſible kinds of it which contain no bitumen, and burn without flame. Theſe reſemble, ſome of them very preciſely, and all of them in a great degree, the products obtained by the diſtillation of the common bituminous coal ; that is, they conſiſt of charcoal, united to an earthy baſis in different proportions. It is natural therefore to conclude, that this ſubſtance was prepared in the mineral regions by the action of heat, which, in ſome inſtances, has driven off the inflammable part of the coal. That the heat ſhould, in ſome caſes, have done ſo, is not inconſiſtent with the general effect attributed to compreſſion. The conditions neceſſary for retaining the more volatile parts, may not have been preſent every where in the ſame degree,

gree, ſo that the latter, though they could not eſcape, may have been forced from one part of a ſtratum, or body of ſtrata, to another.

31. In confirmation of this it muſt be obſerved, that, as the fixed part of coal is thus found in the bowels of the earth, ſeparate from the volatile or bituminous, ſo, in the neighbourhood of coal ſtrata, the latter is ſometimes found without any mixture of the former. The fountains of naphtha and petroleum are well known ; and Dr Hutton has deſcribed a ſtratum of limeſtone, lying in the centre of a coal country, which is pervaded and tinged by bituminous matter, through its whole maſs, and has, at the ſame time, many cloſe cavities in the heart of it, lined with calcareous ſpar, and containing foſſil pitch, ſometimes in large pieces, ſometimes in hemiſpherical drops, ſcattered over the ſurface of the cavities. This combination could only be effected by a part of the inflammable matter of the beds of coal underneath, being driven off by heat, and made to penetrate the limeſtone, while it was yet ſoft and pervious to heated vapours *.

32. Hitherto we have enumerated thoſe foſſils that are either not at all, or very ſparingly ſoluble in water. There are, however, ſaline

* Note IX.

bodies among the mineral ſtrata, ſuch for inſtance as rock-ſalt, which are readily diſſolved in water; and it yet remains to examine by what cauſe their conſolidation has been effected.

Here the theoriſts who conſider water as the ſole agent in the mineralization of foſſils, are indeed delivered from one difficulty, but it is only that they may be harder preſſed on by another. It cannot now be ſaid, that the menſtruum which they employ is incapable of diſſolving the ſubſtances expoſed to its action, as in the caſe of metallic or ſtony bodies; but it may very well be aſked, how the water came to depoſite the ſalts which it held in ſolution, and to depoſite them ſo copiouſly as it has done in many places, without any veſtige of ſimilar depoſition in the places immediately contiguous. If they refuſe to call to their aſſiſtance any other than their favourite element, they will not find it eaſy to anſwer this queſtion, and muſt feel the embarraſſment of a ſyſtem, ſubject to two difficulties, ſo nicely, but ſo unhappily adjuſted, that one of them is always prepared to act whenever the other is removed. If, on the other hand, they will admit the operation of ſubterraneous heat, it appears poſſible, that the local application of ſuch heat

heat may have driven the water, in vapour, from one place to another, and by ſuch action often repeated in the ſame ſpot, may have produced thoſe great accumulations of ſaline matter, that are actually found in the bowels of the earth.

33. But granting that, either in the way juſt pointed out, or in ſome other that is unknown, the ſalt and the water have been ſeparated, ſome further action of heat ſeems requiſite, before a compact, and highly indurated body, like rock-ſalt, could be produced. The mere precipitation of the ſalt, would, as Dr Hutton has obſerved, form only an aſſemblage of looſe cryſtals at the bottom of the ſea, without ſolidity or coheſion: and to convert ſuch a maſs into a firm and ſolid rock, would require the application of ſuch heat as was able to reduce it into fuſion. The conſolidation of rock-ſalt, therefore, however its ſeparation from the water is accounted for, cannot be explained but on the hypotheſis of ſubterraneous heat.

34. Some other phenomena that have been obſerved in ſalt mines, come in ſupport of the ſame concluſion. The ſalt rock of Cheſhire, which lies in thick beds, interpoſed between ſtrata of an argillaceous or marly ſtone, and is itſelf mixed with a conſiderable portion of the ſame earth, exhibits a very great peculiarity in its ſtructure. Though it forms a maſs extreme-

 ly

ly compact, the salt is found to be arranged in round masses of five or six feet in diameter, not truly spherical, but each compressed by those that surround it, so as to have the shape of an irregular polyhedron. These are formed of concentric coats, distinguishable from one another by their colour, that is, probably by the greater or less quantity of earth which they contain, so that the roof of the mine, as it exhibits a horizontal section of them, is divided into polygonal figures, each with a multitude of polygons within it, having altogether no inconsiderable resemblance to a *mosaic* pavement. In the triangular spaces without the polygons, the salt is in coats parallel to the sides of the polygons.

The circumstances which gave rise to this singular structure we should in vain endeavour to define; yet some general conclusions concerning them seem to be within our reach. It is clear that the whole mass of salt was fluid at once, and that the forces, whatever they were, which gave solidity to it, and produced the new arrangement of its particles, were all in action at the same time. The uniformity of the coated structure is a proof of this, and, above all, the compression of the polyhedra, which is always mutual, the flat side of one being turned to the flat side of another, and never an angle to an angle, nor an angle to a side. The coats formed as it were

were round ſo many different centres of attraction, is alſo an appearance quite inconſiſtent with the notion of depoſition; both theſe, however, are compatible with the notion of ſolidity acquired by the refrigeration of a fluid, where the whole maſs is acted on at the ſame time, and where no ſolvent remains to be diſpoſed of after the induration of the reſt.

35. Another ſpecies of foſſil-ſalt exhibits appearances equally favourable to the theory of igneous conſolidation. This is the trona of Africa, which is no other than ſoda, or mineral alkali, in a particular ſtate. The ſpecimen of this foſſil in Dr Black's, now Dr Hope's, collection, is of a ſparry and radiated ſtructure, and is evidently part of the contents of a vein, having a ſtony cruſt adhering to it, on one ſide, with its own ſparry ſtructure complete, on the oppoſite. It contains but about one-ſixth of the water of cryſtallization eſſential to this ſalt when obtained in the humid way; and, what is particularly to be remarked, it does not loſe this water, nor become covered with a powder, like the common alkali, by ſimple expoſure to the air. It is evident, therefore, that this foſſil does not originate from mere precipitation; and when we add, that in its ſparry ſtructure it contains evident marks of having once been fluid, we have

little reaſon to entertain much doubt concerning the principle of its conſolidation.

Thus, then, the teſtimony given to the operation of fire, or heat, as the conſolidating power of the mineral kingdom, is not confined to a few foſſils, but is general over all the ſtrata. How far the unſtratified foſſils agree in ſupporting the ſame concluſion, will be afterwards examined.

3. *Poſition of the Strata* *.

36. We have ſeen of what materials the ſtrata are compoſed, and by what power they have been conſolidated; we are next to inquire, from what cauſe it proceeds, that they are now ſo far removed from the region which they originally occupied, and wherefore, from being all covered by the ocean, they are at preſent raiſed in many places fifteen thouſand feet above its ſurface. Whether this great change of relative place can be beſt accounted for by the depreſſion of the ſea, or the elevation of the ſtrata themſelves, remains to be conſidered.

Of

* Theory of the Earth, vol. i. p. 120.

Of thefe two fuppofitions, the former, at firft fight, feems undoubtedly the moft probable, and we feel lefs reluctance to fuppofe, that a fluid, fo unftable as the ocean, has undergone the great revolution here referred to, than that the folid foundations of the land have moved a fingle fathom from their place. This, however is a mere illufion. Such a depreffion of the level of the fea as is here fuppofed, could not happen without a change proportionally great in the folid part of the globe ; and, though admitted as true, will be found very inadequate to explain the prefent condition of the ftrata.

37. Suppofing the appearances which clearly indicate fubmerfion under water to reach no higher than ten thoufand feet above the prefent level of the fea, and of courfe the furface of the fea to have been formerly higher by that quantity than it is now ; it neceffarily follows, that a bulk of water has difappeared, equal to more than a feven-hundredth part of the whole magnitude of the globe *. The exiftence of empty caverns, of extent fufficient to contain this vaft body of water, and of fuch a convulfion as to lay them open, and give room to the retreat of the fea, are fuppofitions which a philofopher could only be juftified in admitting, if they promifed to furnifh a very complete explanation of appearances.

* Note x.

appearances. But this juſtification is entirely wanting in the preſent caſe; for the retreat of the ocean to a lower level, furniſhes a very partial and imperfect explanation of the phenomena of geology. It will not explain the numberleſs remains of ancient continents that are involved, as we have ſeen, in the preſent, unleſs it be ſuppoſed that the ancient ocean, though it roſe to ſo great a height, had nevertheleſs its ſhores, and was the boundary of land ſtill higher than itſelf. And, as to that which is now more immediately the object of inquiry, the poſition of the ſtrata, though the above hypotheſis would account in ſome ſort for the change of their place, relatively to the level of the ſea; yet, if it ſhall be proved, that the ſtrata have changed their place relatively to each other, and relatively to the plane of the horizon, ſo as to have had an angular motion impreſſed on them, it is evident that, for theſe facts, the retreat of the ſea does not afford even the ſhadow of a theory.

38. Now, it is certain, that many of the ſtrata have been moved angularly, becauſe that, in their original poſition they muſt have been all nearly horizontal. Looſe materials, ſuch as ſand and gravel ſubſiding at the bottom of the ſea, and having their interſtices filled with water, poſſeſs a kind of fluidity: they are diſpoſed to

yield

yield on the ſide oppoſite to that where the preſſure is greateſt, and are therefore, in ſome degree, ſubject to the laws of hydroſtatics. On this account they will arrange themſelves in horizontal layers; and the vibrations of the incumbent fluid, by impreſſing a ſlight motion backward, and forward, on the materials of theſe layers, will very much aſſiſt the accuracy of their level.

It is not, however, meant to deny, that the form of the bottom might influence, in a certain degree, the ſtratification of the ſubſtances depoſited on it. The figure of the lower beds depoſited on an uneven ſurface, would neceſſarily be affected by two cauſes; the inclination of that ſurface, on the one hand, and the tendency to horizontality, on the other; but, as the former cauſe would grow leſs powerful as the diſtance from the bottom increaſed, the latter cauſe would finally prevail, ſo that the upper beds would approach to horizontality, and the lower would neither be exactly parallel to them, nor to one another. Whenever, therefore, we meet with rocks, diſpoſed in layers quite parallel to one another, we may reſt aſſured, that the inequalities of the bottom have had no effect, and that no cauſe has interrupted the ſtatical tendency above explained.

Now,

Now, rocks having their layers exactly parallel, are very common, and prove their original horizontality to have been more precise than we could venture to conclude from analogy alone. In beds of sand-stone, for instance, nothing is more frequent than to see the thin layers of sand, separated from one another by layers still finer of coaly, or micaceous matter, that are almost exactly parallel, and continue so to a great extent without any sensible deviation. These planes can have acquired their parallelism only in consequence of the property of water just stated, by which it renders the surfaces of the layers, which it deposites, parallel to its own surface, and therefore parallel to one another. Though such strata, therefore, may not now be horizontal, they must have been so originally; otherwise it is impossible to discover any cause for their parallelism, or any rule by which it can have been produced.

39. This argument for the original horizontality of the strata, is applicable to those that are now farthest removed from that position. Among such, for instance, that are highly inclined, or even quite vertical, and among those that are bent and incurvated in the most fantastical manner, as happens more especially in the

the primary ſchiſti, we obſerve, through all their ſinuoſities and inflections, an equality of thickneſs and of diſtance among their component laminæ. This equality could only be produced by thoſe laminæ having been originally ſpread out on a flat and level ſurface, from which ſituation, therefore, they muſt afterwards have been lifted up by the action of ſome powerful cauſe, and muſt have ſuffered this diſturbance while they were yet in a certain degree flexible and ductile. Though the primary direction of the force which thus elevated them muſt have been from below upwards, yet it has been ſo combined with the gravity and reſiſtance of the maſs to which it was applied, as to create a lateral and oblique thruſt, and to produce thoſe contortions of the ſtrata, which, when on the great ſcale, are among the moſt ſtriking and inſtructive phenomena of geology.

40. Great additional force is given to this argument, in many caſes, by the nature of the materials of which the ſtratified rocks are compoſed. The beds of breccia and pudding-ſtone, for inſtance, are often in planes almoſt vertical, and at the ſame time contain gravel-ſtones, and other fragments of rock, of ſuch a ſize and weight, that they could not remain in their preſent

ſent poſition an inſtant, if the cement which unites them were to become ſoft; and therefore they certainly had not that poſition at the time when this cement was actually ſoft. This remark has been made by mineralogiſts who were not led to it by any ſyſtem. The judicious and indefatigable obſerver of the Alps, deſcribing the pudding-ſtone of Valorſine, near the ſources of the Arve, tells us, that he was aſtoniſhed to find it in beds almoſt vertical, a ſituation in which it could not poſſibly have been formed. "That particles," he adds, "of extreme tenuity, ſuſpended in a fluid, might become agglutinated, and form vertical beds, is a thing that may be conceived; but that pieces of ſtone, of ſeveral pounds weight, ſhould have reſted on the ſide of a perpendicular wall, till they were enveloped in a ſtony cement, and united into one maſs, is a ſuppoſition impoſſible and abſurd. It ſhould be conſidered, therefore, as a thing demonſtrated, that this pudding-ſtone was formed in a horizontal poſition, or one nearly ſuch, and elevated after its induration. We know not," he continues, "the force by which this elevation has been effected; but it is an important ſtep among the prodigious number of vertical beds that are to be met with in the Alps, to have found ſome

that

that muſt certainly have been formed in a horizontal ſituation*."

41. Nothing can be more ſound and concluſive than this reaſoning; and, had the ingenious author purſued it more ſyſtematically, it muſt have led him to a theory of mountains very little different from that which we are now endeavouring to explain. If ſome of the vertical ſtrata are proved to have been formed horizontally, there can be no reaſon for not extending the ſame concluſion to them all, even if we had not the ſupport of the argument from the parallelifm of the layers, which has been already ſtated.

42. The highly inclined poſition, and the manifold inflexions of the ſtrata, are not the only proofs of the diſturbance that they have ſuffered, and of the violence with which they have been forced up from their original place. Thoſe interruptions of their continuity which are obſerved, both at the ſurface and under it, are evidences of the ſame fact. It is plain, that if they remained now in the ſituation in which they were at firſt depoſited, they would never appear to be ſuddenly broken off. No ſtratum would terminate abruptly; but, however its nature

* Voyages aux Alpes, tom. ii. § 690.

ture and properties might change, it would conſtitute an entire and continued rock, at leaſt where the effects of waſte and *detritus* had not produced a ſeparation. This, however, is very far from being the actual condition of ſtratified bodies. Thoſe that are much inclined, or that make conſiderable angles with the horizontal plane, muſt terminate abruptly where they come up to the ſurface. Their doing ſo is a neceſſary conſequence of their poſition, and furniſhes no argument, it may be ſaid, for their having been diſturbed, different from that which has been already deduced from their inclination. There are, however, inſtances of a breach of continuity in the ſtrata, under the ſurface, that afford a proof of the violence with which they have been diſplaced, different from any hitherto mentioned. Of this nature are the *ſlips* or *ſhifts*, that ſo often perplex the miner in his ſubterraneous journey, and which change at once all thoſe lines and bearings that had hitherto directed his courſe. When his mine reaches a certain plane, which is ſometimes perpendicular, ſometimes oblique to the horizon, he finds the beds of rock broken aſunder, thoſe on the one ſide of the plane having changed their place, by ſliding in a particular direction along the face of the others. In this motion they have ſometimes preſerved their paralleliſm, that is, the ſtrata on on one ſide of the *ſlip* continue parallel to thoſe on the other; in other caſes, the ſtrata on each ſide become inclined to one another, though their identity is ſtill to be recogniſed by their poſſeſſing the ſame thickneſs, and the ſame internal characters. Theſe *ſhifts* are often of great extent, and muſt be meaſured by the quantity of the rock moved, taken in conjunction with the diſtance to which it has been carried. In ſome inſtances, a vein is formed at the plane of the ſhift or ſlip, filled with materials of the kinds which will be hereafter mentioned; in other inſtances, the oppoſite ſides of the rock remain contiguous, or have the interval between them filled with ſoft and unconſolidated earth. All theſe are the undeniable effects of ſome great convulſion, which has ſhaken the very foundations of the earth; but which, far from being a diſorder in nature, is part of a regular ſyſtem, eſſential to the conſtitution and economy of the globe.

The production of the appearances now deſcribed, belongs, without doubt, to different periods of time; and, where ſlips interſect one another, we can often diſtinguiſh the leſs from the more ancient. They are all, however, of a date poſterior to that at which the waving and undulated forms of the ſtrata were acquired, as they do not carry with them any marks of

 the

the ſoftneſs of the rock, but many of its complete induration.

The ſame phenomenon which is thus exemplified on a great ſcale in the bowels of the earth, is often moſt beautifully exhibited in ſingle ſpecimens of ſtone, and is accompanied with this remarkable circumſtance, that the *integrity* of the ſtone is not deſtroyed by the ſhifts, whatever wounds had been made in it being healed, and the parts firmly re-united to one another *.

43. Though ſuch marks of violence as have been now enumerated are common in ſome degree to all the ſtrata, they abound moſt among the primary, and point out theſe as the part of our globe which has been expoſed to the greateſt viciſſitudes. At their junction with the ſecondary, or where they emerge, as it were, from under the latter, phenomena occur, which mark ſome of thoſe viciſſitudes with aſtoniſhing preciſion; phenomena of which the nature was firſt accurately explored, and the conſequences fully deduced, by the geologiſt whoſe ſyſtem I am endeavouring to explain. He obſerved, in ſeveral inſtances, that where the primary ſchiſtus riſes in beds almoſt vertical, it is covered by horizontal layers of ſecondary ſandſtone, which laſt are penetrated by the irregular

* Note XI.

gular tops of the ſchiſtus, and alſo involve fragments of that rock, ſome angular, others round and ſmooth, as if worn by attrition. From this he concluded, that the primary ſtrata, after being formed at the bottom of the ſea, in planes nearly horizontal, were raiſed, ſo as to become almoſt vertical, while they were yet covered by the ocean, and before the ſecondary ſtrata had begun to be depoſited on them. He alſo argued, that, as the fragments of the primary rock, included in the ſecondary, are many of them rounded and worn, the depoſition of the latter muſt have been ſeparated from the elevation of the former by ſuch an interval of time, as gave room for the action of waſte and decay, allowing thoſe fragments firſt to be detached, and afterwards wrought into a round figure *.

44. Indeed, the interpoſition of a breccia between the primary and ſecondary ſtrata, in which the fragments, whether round or angular, are always of the primary rock, is a fact ſo general, and the quantity of this breccia is often ſo great, that it leads to a concluſion more paradoxical than any of the preceding, but from which, nevertheleſs, it ſeems very difficult to with-hold aſſent. Round gravel, when in great abundance, agreeably to a remark already made, muſt neceſſarily be conſidered as a production peculiar

* Note xii.

peculiar to the beds of rivers, or the ſhores of continents, and as hardly ever formed at great depths under the ſurface of the ſea. It ſhould ſeem, then, that the primary ſchiſtus, after attaining its erect poſition, had been raiſed up to the ſurface, where this gravel was formed; and from thence had been let down again to the depths of the ocean, where the ſecondary ſtrata were depoſited on it. Such alternate elevations and depreſſions of the bottom of the ſea, however extraordinary they may ſeem, will appear to make a part of the ſyſtem of the mineral kingdom, from other phenomena hereafter to be deſcribed.

45. On the whole, therefore, by comparing the actual poſition of the ſtrata, their erectneſs, their curvature, the interruptions of their continuity, and the tranſverſe ſtratification of the ſecondary in reſpect of the primary, with the regular and level ſituation which the ſame ſtrata muſt have originally poſſeſſed, we have a complete demonſtration of their having been diſturbed, torn aſunder, and moved angularly, by a force that has, in general, been directed from below upwards. In eſtabliſhing this concluſion, we have reaſoned more from the facts which relate to the *angular elevation* of the ſtrata, than from thoſe which relate to their *abſolute elevation*, or their tranſlation to a greater

diſtance

diſtance from the centre of the earth. This has been done, becauſe the appearances, which reſpect the abſolute lifting up of the ſtrata are more ambiguous than thoſe, which reſpect the change of their angular poſition. The former might be accounted for, could they be ſeparated from the latter, in two ways, viz. either by the retreat of the ſea, or the raiſing up of the land; but the latter can be explained only in one way, and force us of neceſſity to acknowledge the exiſtence of an expanding power, which has acted on the ſtrata with incredible energy, and has been directed from the centre toward the circumference.

46. When we are aſſured of the exiſtence of ſuch a power as this in the mineral regions, we ſhould argue with ſingular inconſiſtency if we did not aſcribe to it all the other appearances of motion in thoſe regions, which it is adequate to produce. If nature in her ſubterraneous abodes is provided with a force that could burſt aſunder the maſſy pavement of the globe, and place the fragments upright upon their edges, could ſhe not, by the ſame effort, raiſe them from the greateſt depths of the ſea, to the higheſt elevation of the land? The cauſe that is adequate to one of theſe effects, is adequate to them both together; for it is a principle well known in mechanical philoſophy, that the force which

 produces

produces a parallel motion, may, according to the way in which it is applied, produce alſo an angular motion, without any diminution of the former effect. It would, therefore, be extremely unphiloſophical to ſuppoſe, that any other cauſe has changed the relative level of the ſtrata, and the ſurface of the ſea, than that which has, in ſo many caſes, raiſed the ſtrata from a horizontal to a highly inclined, or even vertical ſituation: it would be to introduce the action of more cauſes than the phenomena require, and to forget, that nature, whoſe operations we are endeavouring to trace, combines the poſſeſſion of infinite reſources with the moſt economical application of them.

47. From all, therefore, that relates to the poſition of the ſtrata, I think I am juſtified in affirming, that their diſturbance and removal from the place of their original formation, by a force directed from below upwards, is a fact in the natural hiſtory of the earth, as perfectly aſcertained as any thing which is not the ſubject of immediate obſervation. As to the power by which this great effect has been produced, we cannot expect to decide with equal evidence, but muſt be contented to paſs from what is certain to what is probable. We may, then, remark, that of the forces in nature to which our experience does in any degree extend,

tend, none ſeems ſo capable of the effect we would aſcribe to it, as the expanſive power of heat; a power to which no limits can be ſet, and one, which, on grounds quite independent of the elevation of the ſtrata, has been already concluded to act with great energy in the ſubterraneous regions. We have, indeed, no other alternative, but either to adopt this explanation, or to aſcribe the facts in queſtion to ſome ſecret and unknown cauſe, though we are ignorant of its nature, and have no evidence of its exiſtence.

We are therefore to ſuppoſe, that the power of the ſame ſubterraneous heat, which conſolidated and mineralized the ſtrata at the bottom of the ſea, has ſince raiſed them up to the height at which they are now placed, and has given them the various inclinations to the horizon which they are found actually to poſſeſs.

48. The probability of this hypotheſis will be greatly increaſed, when it is conſidered, that, beſides thoſe now enumerated, there are other indications of movement among the bodies of the mineral kingdom, where effects of heat more characteriſtic than ſimple expanſion are clearly to be diſcovered. Thus, on examining the marks of diſorder and movement which are found among the ſtrata, it cannot fail to be obſerved, that not-

withſtanding the fracture and diſlocation, of which they afford ſo many examples, there are few empty ſpaces to be met with among them, as far as our obſervation extends. The breaches and ſeparations are numerous, and diſtinct; but they are, for the moſt part, completely filled up with minerals of a kind quite different from the rock on each ſide of them, and remarkable for containing no veſtiges of ſtratification. We are thus led to conſider the unſtratified foſſils, the ſecond of the diviſions into which the whole mineral kingdom, viewed geologically, ought to be diſtinguiſhed. Theſe foſſils are immediately connected with the diſturbance of the ſtrata, and appear, in many inſtances, to have been the inſtruments of their elevation.

SECTION II.

OF THE PHENOMENA PECULIAR TO UNSTRATIFIED BODIES.

1. *Metallic Veins.*

49. THE unſtratified minerals exiſt either in veins, interſecting the ſtratified, or in maſſes ſurrounded by them. Veins are of various kinds, and may in general be defined, ſeparations in the continuity of a rock, of a determinate width, but extending indefinitely in length and depth, and filled with mineral ſubſtances, different from the rock itſelf. The mineral veins, ſtrictly ſo called, are thoſe filled with ſparry or cryſtallized ſubſtances, and containing the metallic ores.

That theſe veins are of a formation ſubſequent to the hardening and conſolidation of the ſtrata which they traverſe, is too obvious to require any proof; and it is no leſs clear, from the cryſtallized and ſparry ſtructure of the ſubſtances contained in them, that theſe ſubſtances muſt have concreted from a fluid ſtate. Now, that this fluidity was ſimple, like that of fuſion by heat, and not compound, like that of ſolution in a menſtruum, is inferred from many phenomena. It is inferred from the acknowledged inſolubility

inſolubility of the ſubſtances that fill the veins, in any one menſtruum whatſoever; from the total diſappearance of the ſolvent, if there was any; from the complete filling up of the vein by the ſubſtances which that ſolvent had depoſited; from the entire abſence of all the appearances of horizontal or gradual depoſition; and, laſtly, from the exiſtence of cloſe cavities, lined with cryſtals, and admitting no egreſs to any thing but heat.

50. To the ſame effect may be mentioned thoſe groups of cryſtals compoſed of ſubſtances the moſt different, that are united in the ſame ſpecimen, all interſecting and mutually impreſſing one another. Theſe admit of being explained, on the ſuppoſition that they were originally in fuſion, and became ſolid by the loſs of heat; a cauſe that acted on them all alike, and alike impelled them to cryſtallize: But the appearances of ſimultaneous cryſtallization ſeem incompatible with the nature of depoſition from a ſolvent, where, with reſpect to different ſubſtances, the effects muſt take place ſlowly, and in ſucceſſion.

51. The metals contained in the veins which we are now treating of, appear very commonly in the form of an ore, mineralized by ſulphur. Their union with this latter ſubſtance can be produced, as we know, by heat, but hardly by the way of ſolution in a menſtruum, and certainly

tainly not at all, if that menſtruum is nothing elſe than water. The metals, therefore, when mineralized by ſulphur, give no countenance to the hypotheſis of aqueous ſolution; and ſtill leſs do they give any when they are found native, as it is called, that is, malleable, pure and uncombined with any other ſubſtance. The great maſſes of native iron found in Siberia and South America are well known; and nothing certainly can leſs reſemble the products of a chemical precipitation. Gold, however, the moſt perfect of the metals, is found native moſt frequently; the others more rarely, in proportion nearly to the facility of their combination with ſulphur. Of all ſuch ſpecimens it may be ſafely affirmed, that if they have ever been fluid, or even ſoft, they muſt have been ſo by the action of heat; for, to ſuppoſe that a metal has been precipitated, pure and uncombined from any menſtruum, is to treſpaſs againſt all analogy, and to maintain a phyſical impoſſibility. But it is certain, that many of the native metals have once been in a ſtate of ſoftneſs, becauſe they bear on them impreſſions which they could not have received but when they were ſoft. Thus, gold is often impreſſed by quartz and other ſtones, which ſtill adhere to it, or are involved in it. Specimens of quartz, containing gold and ſilver ſhooting through them,

them, with the moſt beautiful and varied ramifications, are every where to be met with in the cabinets of the curious; and contain, in their ſtructure, the cleareſt proof, that the metal and the quartz have been both ſoft, and have cryſtallized together. By the compactneſs, alſo, of the body which they form, they ſhow, that when they acquired ſolidity, it was by the concretion of the whole maſs, and not by ſuch partial concretion as takes place when a ſolvent is ſeparated from ſubſtances which it held in ſolution.

52. Native copper is very abundant; and ſome ſpecimens of it have been found cryſtallized. Here the cryſtallization of the metal is a proof that it has paſſed from a fluid to a ſolid ſtate; and its purity is a proof that it did not make that tranſition by being precipitated from a menſtruum.

53. Again, pieces of native manganeſe have been found poſſeſſing ſo exactly the characters peculiar to that metal when reduced in our furnaces, that it is impoſſible to conſider them as deriving their figure and ſolidity from any cauſe but fuſion. The ingenious author who deſcribes theſe ſpecimens, La Peyrouſe, was ſo forcibly ſtruck with this reſemblance, that he immediately drew the ſame concluſion from it which is drawn here, attributing the only difference,

ence, which he remarked between the native and the artificial *regulus*, to the different energy with which the ſame agent works when employed by nature and by art *.

54. All theſe appearances conſpire to prove, that the materials which fill the mineral veins were melted by heat, and forcibly injected, in that ſtate, into the clefts and fiſſures of the ſtrata. Theſe fiſſures we muſt conceive to have ariſen, not merely from the ſhrinking of the ſtrata while they acquired hardneſs and ſolidity, but from the violence done to them, when they were heaved up and elevated in the manner which has already been explained †.

55. When theſe ſuppoſitions are once admitted, the other leading facts in the hiſtory of metallic veins will be readily accounted for. Thus, for inſtance, it is evident to what we muſt aſcribe the fragments of the ſurrounding rock that are often found immerſed in the veins, and encompaſſed on all ſides by cryſtallized ſubſtances. Theſe fragments being no doubt detached by the concuſſion, which at once tore aſunder and elevated the ſtrata, were ſuſtained by the melted matter that flowed at the ſame time upward

* Theory of the Earth, vol. i. p. 68. Journal de Phyſ. Janvier 1786.

† Note xiii.

ward through the vein. Large maſſes of rock are often found in this manner completely inſulated; one of theſe, which M. de Luc has deſcribed with great accuracy, is no leſs than a vaſt ſegment of a mountain *.

56. The immenſe violence which has accompanied the formation of mineral veins, is particularly marked by the ſlips and ſhifts of the ſtrata on each ſide of them, all tending to ſhow what mighty changes have taken place in thoſe regions, which our imagination erroneouſly paints as the abode of everlaſting ſilence and reſt. This ſhifting of the ſtrata is beſt obſerved, where the veins make a tranſverſe ſection of beds of rock, conſiderably inclined to the horizon. There it is common to ſee the beds on one ſide of the vein ſlipped along from the correſponding beds on the other ſide, and removed ſometimes in a horizontal, ſometimes in an oblique direction. In this way, not only the ſtrata are ſhifted, but veins, which interſect one another, are alſo ſhifted themſelves. They are *heaved*, as it is called in the ſignificant language of the miners, and forced out of their direction. It is impoſſible, in ſuch a caſe, but to connect in the mind the formation of

* Lettres Phyſiques, &c. tom. iii. p. 361.

of the vein, and the production of the ſlips which accompany it, and to regard them as parts of the ſame phenomenon.

57. Where theſe ſlips are horizontal, and exhibit great bodies of ſtrata carried from their place, while the parts of the transferred maſs remain undiſturbed relatively to one another, they furniſh a clear proof, that this change of place has not ariſen from the falling in of the roofs of caverns, as ſome geologiſts ſuppoſe. The horizontal direction, and the regularity of the movement, are incompatible with the action of ſuch a cauſe as this; and indeed it is highly intereſting to remark, in the midſt of the ſigns of diſturbance which prevail in the bowels of the earth, that there reigns a certain ſymmetry and order, which indicate the action of a force of incredible magnitude, but ſlow and gradual in its effects. The parts of the maſs moved are undiſturbed relatively to one another: what has been broken has been cemented: the breaches of continuity have been filled up and healed; and every where we ſee the operation of a cauſe that could unite as well as ſeparate. The twofold action of heat to expand and to melt, could ſcarce be pointed out more clearly by any ſyſtem of appearances.

58. As a long period was no doubt required for the elevation of the ſtrata, the rents made

in

in them are not all of the ſame date, nor the veins all of the ſame formation. This is clear in the caſe of one vein producing a ſhift or ſlip in another; for the vein which forces the other out of its place, and preſerves its own direction, is evidently the more recent of the two, and muſt have had its materials in a ſtate of activity, when thoſe of the other were inert. Sometimes, alſo, at the interſection of two veins, we may trace the current of the materials of the one, acroſs thoſe of the other; and here, of conſequence, the relative antiquity is determined juſt as in the former inſtance.

59. The want of any appearance of ſtratification in mineral veins has already been taken notice of. There is, however, to be obſerved, in many inſtances, a tendency to a regular arrangement of the ſubſtances contained in them; thoſe of the ſame kind forming coats parallel to the ſides of the vein, and nearly of an equal thickneſs. This phenomenon is conſidered as one of the ſtrongeſt arguments in favour of the Neptunian ſyſtem, but has nothing in it, in the leaſt incompatible with that theory which aſcribes the formation of veins to the action of ſubterraneous heat. When melted matter from the mineral regions was thrown up into the veins, that which was neareſt to the ſides would ſooneſt loſe its heat. The ſimilar ſubſtances, alſo, would unite while

this

this procefs was going forward, and would cryftallize, as in other cafes of congelation, from the fides toward the interior. There is the more reafon for fuppofing this to have been the cafe, that the fame fort of coating is often obferved on the infide of clofe cavities, which are, neverthelefs, fo conftructed, as to afford a demonftration that no chemical folvent was ever included in them, (§ 74.). Some veins, it muft alfo be confidered, may have been filled by fucceffive injections of melted matter, and this would naturally give rife to a variety of feparate incruftations *.

60. In the view now given of metallic veins, they have been confidered as traverfing only the ftratified parts of the globe. They do, however, occafionally interfect the unftratified parts, particularly the granite, the fame vein often continuing its courfe acrofs rocks of both kinds, without fuffering any material change; and, if we have hitherto paid no attention to this circumftance, it is becaufe the order purfued in this effay required, that the relation of the veins to ftratified bodies fhould be firft treated of. Befides, the facts in the natural hiftory of veins, whether contained in ftratified or unftratified

 rocks,

* See fome farther remarks on this fubject at NOTE XIII.

rocks, are ſo nearly alike, that in a general view of geology, they do not require to be diſtinguiſhed. It is material to remark, that, though metallic veins are found indiſcriminately in all the different kinds of rock, whether ſtratified or otherwiſe, they are moſt abundant in the claſs of primary ſchiſti. All the countries moſt remarkable for their mines, and the mountains diſtinguiſhed by the name of metalliferous, are primary, and the inſtance of Derbyſhire is perhaps the moſt conſiderable exception to this rule, that is known. This preference, which the metals appear to give to the primary ſtrata, is very conſiſtent with Dr Hutton's theory, which repreſents the rocks of that order as being moſt changed from their original poſition, and thoſe on which the diſturbing forces of the ſubterraneous regions have acted moſt frequently, and with greateſt energy. The primary ſtrata are the loweſt, alſo, and have the moſt direct communication with thoſe regions from which the mineral veins derive all their riches.

2. *Of Whinſtone.*

61. Beſide the veins filled with ſpar, and containing the metallic ores, the ſtrata are interſected by veins of whinſtone, porphyry and granite, the

the characters of which are next to be examined.

The term *whin*, or *whinstone*, with Dr Hutton, like the word *trap*, with the German mineralogists, denotes a class of stones, comprehending several distinct species, or at least varieties. The common *basalt*, the *wacken*, *mullen*, and *crag* of Kirwan, the *grünstein* of Werner, and the *amygdaloid*, are comprehended under the name of whin. All these stones have a tendency to a spathose structure, and discover at least the rudiments of crystallization. They are, at the same time, without any mark of stratification in their internal texture, as they are also, for the most part, in their outward configuration; and, as the different species here enumerated compose, not unfrequently, parts of the same continuous rock, the change from one to another being made through a series of insensible gradations, they may safely be regarded by the geologist as belonging to the same *genus*.

62. Whin, though not stratified, exists in two different ways, that is, either in veins, (called in Scotland *dykes*), traversing the strata like the veins already described, or in irregular masses, incumbent on the strata, and sometimes interposed between them. In both these forms, whinstone has nearly the same characters, and

bears, in all its varieties, a moſt ſtriking reſemblance to the lavas which have actually flowed from volcanoes on the ſurface of the earth. This reſemblance is ſo great, that the two ſubſtances have been often miſtaken for one another; and many rocks, which have been pronounced to be the remains of extinguiſhed volcanoes, by mineralogiſts of no inconſiderable name, have been found, on cloſer examination, to be nothing elſe than maſſes or veins of whinſtone. This latter ſtone is indeed only to be diſtinguiſhed from the former, by a careful examination of the internal characters of both; and chiefly from this circumſtance, that whinſtone often contains calcareous ſpar and zeolite, whereas neither of theſe ſubſtances is found in ſuch lavas, as are certainly known to have been thrown out by volcanic exploſions.

Now, from theſe circumſtances of affinity between lava and whinſtone, on the one hand, and of diverſity on the other, as the formation of the one is known, it ſhould ſeem that ſome probable concluſion may be drawn concerning the formation of the other. The affinity in queſtion is conſtant and eſſential; the difference variable and accidental; and this naturally leads to ſuſpect, that the two ſtones have the ſame origin; and that, as lava is certainly a production of fire, ſo probably is whinſtone.

63. But

63. But, in order to ſee whether this hypotheſis will explain the diverſity of the two ſubſtances, without which it will not be entitled to much attention, we muſt remark, that the preſence of carbonat of lime in a body that has been fuſed, argues, agreeably to the principles formerly explained, that the fuſion was brought about under a great compreſſing force, that is to ſay, deep in the bowels of the earth, or in the great laboratory of the mineral regions. We are, therefore, to ſuppoſe that the fuſion of the whin was performed in thoſe regions, where the compreſſion was ſufficient to preſerve the carbonic gas in union with the calcareous earth, ſo that theſe two ſubſtances melted together, and, on cooling, cryſtallized into ſpar. In the lavas, again, thrown out by volcanic eruption, the fuſion, as we know, wherever it may begin, continues in the open air, where the preſſure is only that of the atmoſphere: the calcareous earth, which, therefore, may have been, in the form of a carbonat, among the materials of this lava, muſt be converted into quicklime, and become infuſible; hence the want of calcareous ſpar in lavas that have flowed at the ſurface.

Thus, whinſtone is to be accounted a ſubterraneous, or *un-erupted* lava; and our theory has the advantage of explaining both the affinity and the difference between thefe ftony bodies, without the introduction of any new hypothefis. In the Neptunian fyftem, the affinity of whinftone and lava is a paradox which admits of no folution.

64. The columnar ftructure fometimes found in that fpecies of whinftone called bafaltes, is a fact which has given rife to much difcuffion; and it muft be confeffed, that though one of the moft ftriking and peculiar characters of this foffil, it is not that which gives the cleareft and moft direct information concerning its origin. One circumftance, however, very much in favour of the opinion that bafaltic rocks owe their formation to fire, is, that the columnar form is fometimes affumed by the lava actually erupted from volcanoes. Now, it is certainly of no fmall importance, to have the fynthetic argument on our fide, and to know, that bafaltic columns can be produced by fire; though, no doubt, to give abfolute certainty to our conclufion, it would be neceffary to fhow, that there are in nature no other means but this by which thefe columns can be formed. This fort of evidence is hardly to be looked for; but fince the power of fufion, to produce the phenomena in queftion, is perfectly eftablifhed, and fince the production of the fame phenomena in the humid way is a

mere

mere hypothefis, if there be the leaft reafon to fufpect the action of fubterraneous heat as one of the caufes of mineralization, every maxim of found philofophy requires that the bafaltic ftructure, in all cafes, fhould be afcribed to it.

65. The Neptunifts will no doubt allege, with BERGMAN, that, in the drying of ftarch, clay, and a few other fubftances, fomething analogous to bafaltic columns is produced. Here, however, a moft important difference is to be remarked, correfponding very exactly to one of the characters which we have all along obferved to diftinguifh the products of aqueous, from thofe of igneous confolidation. The columns formed by the fubftances juft mentioned, are diftant from one another: they are feparated by fiffures which widen from the bottom to the top, and which arife from the fhrinking and drying of the mafs. In the bafaltic columns, no fuch openings, nor vacuity of any kind is found; the pillars are in contact, and, though perfectly diftinct, are fo clofe, that the fharp edge of a wedge can hardly be introduced between them. This is a great peculiarity in the bafaltic ftructure, and is ftrongly expreffive of this fact, that the mafs was all fluid together, and that its parts took their new arrangement, not in confequence of the feparation of a fluid from a folid part, by which great fhrinking and much empty fpace

might be produced; but in consequence of a cause which, like refrigeration, acted equally on all the parts of the mass, and preserved their absolute contact after their fluidity had ceased.

66. A mark of fusion, or at least of the operation of heat, which whinstone possesses in common with many other minerals, is its being penetrated by pyrites, a substance, as has been already remarked, that is of all others most exclusively the production of fire. Another mark of fusion, more distinctive of whin, is, that both in veins and in masses it sometimes includes pieces of sandstone, or of the other contiguous strata, completely insulated, and having the appearance of fragments of rock, floating in a fluid sufficiently dense and ponderous to sustain their weight. Though these fragments have been too refractory to be reduced into fusion themselves, they have not remained entirely unchanged, but are, in general, extremely indurated, in comparison of the rock from which they appear to have been detached.

67. Similar instances of extraordinary induration are observed in the parts of the strata in contact with whinstone, whether they form the sides of the veins, or the floors, and roofs of the masses into which the whinstone is distributed. The strata whether sandy or argillaceous, in such situations, are usually extremely hard and consolidated;

confolidated; the former in particular lofe their granulated texture, and are fometimes converted into perfect jafper. This interefting remark was firft made by Dr Hutton, and the truth of it has been verified by a great number of fubfequent obfervations.

68. To the fame excellent geologift we are indebted for the knowledge of an analogous fact, attendant on the paffage of whinftone veins through coal ftrata. As the beds of ftone where they are in contact with veins of whin, feem to acquire additional induration, fo thofe of coal, in like circumftances, are frequently found to have loft their fufibility, and to be reduced nearly to the condition of coke, or of charcoal. The exiftence of coal of this kind has been already mentioned, and confidered as a proof of the operation of fubterraneous heat. In the inftances here referred to, that is, where the charring of the coal is limited to thofe parts of the ftrata which are in contact with the whin, or in its immediate vicinity, the heat is pointed out as refiding in the vein; and this is to be accounted for only on the fuppofition of the melted whin, at a period fubfequent to the confolidation of the coal, having flowed through the openings of the ftrata. The heat has been powerful enough, in many cafes, to drive off the bituminous matter of the coal, and to force it into colder and more diftant

diſtant parts. Few facts, in the hiſtory of foſſils, are more remarkable than this, and none more directly aſſimilates the operations of the mineral regions, with thoſe that take place at the ſurface of the earth.

69. Again, the diſturbance of the ſtrata, wherever veins of whinſtone abound, if not a direct proof of the original fluidity of the whinſtone, is a clear indication of the violence with which it was introduced into its place. This diſturbance of the poſition of the ſtrata, by ſhifting, unuſual elevation, and other irregularities, where they are interſected by whinſtone veins, is a fact ſo well known to miners, that when they meet with any ſudden change in the lying of the *metals*, they are wont to foretel their approach to maſſes, or veins of unſtratified matter; and, in their figurative language, point them out as the cauſes of the confuſion with which they are ſo generally accompanied *. The mineral veins likewiſe, as well as the ſtrata, are often heaved and ſhifted by the veins of whinſtone.

70. Whinſtone of every ſpecies is found frequently interpoſed in tabular maſſes, between beds of ſtratified rocks; and it then adds to the indications

* A *Trouble* is the name which the colliers in this country give to a vein of whinſtone.

indications of its igneous origin, already enumerated, ſome others that are peculiar to it when in this ſituation. In ſuch inſtances, it is not uncommon to find the ſtrata in ſome places, contiguous to the whin, elevated, and bent with their concavity upward, ſo that they appear clearly to have been acted on by a force that proceeded from below, at the ſame time that they were ſoftened, and rendered in ſome degree flexible: it is needleſs to remark, that theſe effects can be explained by nothing but the fuſion of the whin; and that the great force with which it was impelled againſt the ſtrata, could be produced by no cauſe but heat, acting in the manner that is here ſuppoſed.

71. Again, if it be true that the maſſes of whin, thus interpoſed among the ſtrata, were introduced there, after the formation of the latter, we might expect to find, at leaſt in many inſtances, that the beds on which the whinſtone reſts, and thoſe by which it is covered, are exactly alike. If theſe beds were once contiguous, and have been only heaved up and ſeparated by the irruption of a fluid maſs of ſubterraneous lava, their identity ſhould ſtill be recogniſed. Now, this is preciſely what is obſerved; it is known to hold in a vaſt number of inſtances, and is ſtrikingly exemplified in the rock of *Saliſbury Crag*, near Edinburgh.

This

This ſimilarity of the ſtrata that cover the maſſes of whinſtone, to thoſe that ſerve as the baſe on which they reſt, and again the diſſimilitude of both to the interpoſed maſs, are facts which I think can hardly receive any explanation, on the principles of the Neptunian theory. If theſe rocks, both ſtratified and unſtratified, are to be regarded as productions of the ſea, the circumſtances would require to be pointed out, which have determined the whinſtone, and the beds that are all round it, to be ſo extremely unlike in their ſtructure, though formed at the ſame time, and in the immediate vicinity of one another; as alſo thoſe circumſtances, on the other hand, which determined the ſtratified depoſites above and below the whinſtone, to be preciſely the ſame, though the times of their formation muſt have been very different. The homogeneous ſubſtances, thus, placed at a diſtance, and the heterogeneous brought ſo cloſely together, are phenomena equally unaccountable, in a theory that aſcribes their origin to the operation of the ſame element, and that neceſſarily dates their formation according to the order in which they lie, one above another.

72. If, indeed, in theſe inſtances, the gradation were inſenſible, as ſome have aſſerted it to be, between the ſtrata and the interpoſed maſs, ſo that it was impoſſible to point out the line

where

where the one ended and the other began, whatever difficulties we might perceive in the Neptunian theory, we ſhould find it hard to ſubſtitute a better in its room. But the truth ſeems to be, that, in the caſes we are now treating of, no ſuch gradation exiſts; and that, though where the two kinds of rock come into contact a change is often obſerved, by the ſtrata having acquired an additional degree of induration, yet the line of ſeparation is well defined, and can be preciſely aſcertained. This at leaſt is certain, that innumerable ſpecimens, exhibiting ſuch lines of ſeparation, are to be met with; and wherever care has been taken to obtain a freſh fracture of the ſtone, and to remove the effects of accidental cauſes, even where the two rocks are moſt firmly united, and moſt cloſely aſſimilated, I am perſuaded that no uncertainty has ever remained as to the line of their ſeparation. For theſe reaſons, it ſeems probable that the gradual tranſition of baſaltes into the adjoining ſtrata, is in all caſes imaginary, and is, in truth, a mere illuſion, proceeding from haſty and inaccurate obſervation.

73. Another remarkable fact in the natural hiſtory of the whinſtone rocks, remains yet to be mentioned, and with it I ſhall conclude the argument, as far as theſe rocks are concerned.

Some

Some of the ſpecies of whinſtone are the common matrices of agates and chalcedonies, which lie incloſed in them in the form of round nodules. The original fluidity of theſe nodules is evinced by their figured, and ſometimes cryſtallized ſtructure, and indeed is ſo generally admitted, that the only queſtion concerning them is, whether this fluidity was the effect of heat or of ſolution. To anſwer this queſtion, Dr Hutton obſerves, that the formation of the concentric coats, of which the agate is uſually compoſed, has evidently proceeded from the circumference toward the centre, the exterior coats always impreſſing the interior, but never the reverſe. The ſame thing alſo follows from this other fact, that when there is any vacuity within the agate, it is uſually at the centre, and there too are found the regular cryſtals, when any ſuch have been formed. It therefore appears certain, that the progreſs of conſolidation has been from the circumference inwards, and that the outward coats of the agate were the firſt to acquire ſolidity and hardneſs.

74. Now, it muſt be conſidered that theſe coats are highly conſolidated; that they are of very pure ſiliceous matter, and are utterly impervious to every ſubſtance which we know of, except light and heat. It is plain, therefore, that whatever

ever at any time, during the progreſs of conſolidation, was contained within the coats already formed, muſt have remained there as long as the agate was entire, without the leaſt poſſibility of eſcape. But nothing is found within the coats of the agate ſave its own ſubſtance; therefore no extraneous ſubſtance, that is to ſay no ſolvent, was ever included within them. The fluidity of the agate was therefore ſimple, and unaſſiſted by any menſtruum.

In this argument, nothing appears to me wanting, that is neceſſary to the perfection of a phyſical, I had almoſt ſaid of a mathematical, demonſtration. It ſeems, indeed, to be impoſſible that the igneous origin of foſſils could be recorded in plainer language, than by the phenomenon which has juſt been deſcribed.

75. The examination of particular ſpecimens of agates and chalcedonies, affords many more arguments of the ſame kind, which Dr Hutton uſed to deduce with an acuteneſs and vivacity, which his friends have often liſtened to with great admiration and delight *. Theſe, however, muſt be paſſed over at preſent; and I have only further to remark, that a ſeries of the moſt intereſting experiments, inſtituted by Sir James Hall, and publiſhed in the Tranſactions of the Royal Society of Edinburgh †, has removed the only

* Note xiv.

† Vol. v. p. 43.

only remaining objection that could be urged againſt the igneous origin of whinſtone. This objection is founded on the common obſervation, that when a piece of whinſtone or baſaltes is actually melted in a crucible, on cooling, it becomes glaſs, and loſes its original character entirely; and from thence it was concluded, that this character had not been originally produced by fuſion. The experiments above mentioned, however, have ſhewn, in the moſt ſatisfactory manner, that melted whin, by *regulated* or by ſlow cooling, is prevented from aſſuming the appearance of glaſs, and becomes a ſtony ſubſtance, hardly to be diſtinguiſhed from whinſtone or lava.

The experiments of another ingenious chemiſt, Dr KENNEDY, have ſhewn, that whinſtone contains mineral alkali, by which, of courſe, its fuſion muſt have been aſſiſted *. Dr Hutton uſed to aſcribe its fuſibility, in a great meaſure at leaſt, to the quantity of iron contained in it: both theſe cauſes have no doubt united to render it more eaſily melted than the ordinary materials of the ſtrata.

76. In a word, therefore, to conceive aright the origin of that claſs of unſtratified rocks, diſtinguiſhed by the name of whinſtone, we muſt ſuppoſe,

* Tranſ. R. S. Edin. vol. v. p. 85.

poſe, that long after the conſolidation of the ſtrata, and during the time of their elevation, the materials of the former were melted by the force of ſubterraneous heat, and injected among the rents and fiſſures of the rocks already formed. In this manner were produced the veins or dikes of whinſtone; and, where circumſtances allowed the ſtream of melted matter to diffuſe itſelf more widely, tabular maſſes were formed, which were afterwards raiſed up, together with the ſurrounding ſtrata, above the level of the ſea, and have been ſince laid open by the operation of thoſe cauſes that continually change and waſte the ſurface of the land.

Theſe unſtratified rocks are not, however, all the work of the ſame period; they differ evidently in the date of their formation, and it is not unuſual, to find tabular maſſes of one ſpecies of whin, interſected by veins of another ſpecies. Indeed, of all the foſſil bodies which compoſe the preſent land, the veins of whin appear to be the moſt recently conſolidated *.

Porphyry may ſo properly be regarded as a variety of whin, diſtinguiſhed only by involving cryſtallized feltſpar, that, in a geological ſketch like the preſent, it is hardly entitled to a ſeparate article. Like the other kinds of

 whin,

* Note xiv.

whin, it exifts both in veins and in tabular maffes, having, no doubt, an origin fimilar to that which has juft been defcribed. Porphyry, however, has the peculiarity of being rarely found in any but the primary ftrata; it feems to be the whinftone of the old world, or at leaft that which is of higheft antiquity in the prefent. It no-where, I believe, affumes a columnar, or bafaltic appearance, of any regularity; but this is alfo true of many other varieties of whin, of all, indeed, except the moft compact and homogeneous. Thefe differences are not fo confiderable as to require our entering into any particular detail concerning the natural hiftory of this foffil.

3. *Granite.*

77. The term Granite is ufed by Dr Hutton to fignify an aggregate ftone, in which quartz, feltfpar and mica are found diftinct from one another, and not difpofed in layers. The addition of hornblend, fchorl, or garnet, to the three ingredients juft mentioned, is not underftood to alter the *genus* of the ftone, but only to conftitute a fpecific difference, which it is the bufinefs of lithology to mark by fome appropriate character, annexed to the generic name of granite.

The

The foſſil now defined exiſts, like whinſtone and porphyry, both in maſſes and in veins, though moſt frequently in the former. It is like them unſtratified in its texture, and is regarded here, as being alſo unſtratified in its outward ſtructure *. One ingredient which is eſſential to granite, namely, quartz, is not contained in whinſtone; and this circumſtance ſerves to diſtinguiſh theſe *genera* from one another, though, in other reſpects, they ſeem to be united by a chain of inſenſible gradations, from the

* Thoſe rocks that conſiſt of the ingredients here enumerated, if they have at the ſame time a ſchiſtoſe texture, or a diſpoſition into layers, are properly diſtinguiſhed from granite, and called Gneiſs, or Granitic Schiſtus. But it has been queſtioned whether a ſtone does not exiſt compoſed of theſe ingredients, and deſtitute of a ſchiſtoſe texture, but yet divided into large beds, viſible in its external form. Dr Hutton ſuppoſes ſuch a ſtone not to exiſt, or at leaſt not to conſtitute any ſuch proportion of the mineral kingdom, as to entitle it to particular conſideration, in the general ſpeculations of geology.

Whether this ſuppoſition is perfectly correct, may require to be farther conſidered: this, however, is certain, that a rock, in all reſpects conformable to it, compoſes a great proportion of what are uſually called the granite mountains. See NOTE XV.

moſt homogeneous baſaltes, to granite the moſt highly cryſtallized.

78. Granite, it has been juſt ſaid, exiſts moſt commonly in maſſes; and theſe maſſes are rarely, if ever, incumbent on any other rock: they are the baſis on which others reſt, and ſeem, for the moſt part, to riſe up from under the ancient, or primary ſtrata. The granite, therefore, wherever it is found, is inferior to every other rock; and as it alſo compoſes many of the greateſt mountains, it has the peculiarity of being elevated the higheſt into the atmoſphere, and ſunk the deepeſt under the ſurface, of all the mineral ſubſtances with which we are acquainted.

Notwithſtanding the circumſtance of not being alternated with ſtratified bodies, which conſtitutes a remarkable difference between granite and whinſtone, the affinity of theſe foſſils is ſuch as to make the ſimilarity of their origin by no means improbable. Accordingly, in Dr Hutton's theory, granite is regarded as a ſtone of more recent formation than the ſtrata incumbent on it; as a ſubſtance which has been melted by heat, and which, when forced up from the mineral regions, has elevated the ſtrata at the ſame time.

79. That granite has undergone a change from a fluid to a ſolid ſtate, is evinced from the cryſtallized ſtructure in which ſome of its component

nent parts are ufually found. This cryftallization is particularly to be remarked of the feltfpar, and alfo of the fchorl, where there is any admixture of that fubftance, whether in flender fpiculæ, or in larger maffes. The quartz itfelf is in fome cafes cryftallized, and is fo, perhaps, more frequently than is generally fuppofed. The fluidity of granite, in fome former period of its exiftence, is fo evident from this, as to make it appear fingular that it fhould ever have been confidered as a foffil that had remained always the fame, and one, into the origin of which it was needlefs to inquire. If the regular forms of cryftallization are not to be received as proofs of the fubftance to which they belong having paffed from a fluid to a folid ftate, neither are the figures of fhells and of other fuppofed petrifactions, to be taken as indications of a paffage from the animal to the mineral kingdom; fo that there is an end of all geological theories, and of all reafonings concerning the ancient condition of the globe. To an argument which ftrikes equally at the root of all theories, it belongs not to this, in particular, to make any reply.

80. We fhall, therefore, confider it as admitted, that the materials of the granite were originally fluid; and, in addition to this, we think it can eafily be proved, that this fluidity was

 not

not that of the elements taken ſeparately, but of the entire maſs. This laſt concluſion follows, from the ſtructure of thoſe ſpecimens, where one of the ſubſtances is impreſſed by the forms which are peculiar to another. Thus, in the Portſoy granite *, which Dr Hutton has ſo minutely deſcribed, the quartz is impreſſed by the rhomboidal cryſtals of the feltſpar, and the ſtone thus formed is compact and highly conſolidated. Hence, this granite is not a congeries of parts, which, after being ſeparately formed, were ſomehow brought together and agglutinated; but it is certain that the quartz, at leaſt, was fluid when it was moulded on the feltſpar. In other granites, the impreſſions of the ſubſtances on one another are obſerved in a different order, and the quartz gives its form to the feltſpar. This, however, is more unuſual; the quartz is commonly the ſubſtance which has received the impreſſions of all the reſt; and the ſpiculæ of ſchorl often ſhoot both acroſs it and the feltſpar.

The ingredients of granite were therefore fluid when mixed, or at leaſt when in contact with one another. Now, this fluidity was not the effect of ſolution in a menſtruum; for, in that caſe, one kind of cryſtal ought not to impreſs another, but each of them ſhould have its own peculiar ſhape.

81. The

* Theory of the Earth, vol. i. p. 104.

81. The perfect consolidation of many granites, furnishes an argument to the same effect. For, agreeably to what was already observed, in treating of the strata, a substance, when crystallizing, or passing from a fluid to a solid state, cannot be free from porosity, much less fill up completely a space of a given form, if, at the same time, any solvent is separated from it; because the solvent so separated would still occupy a certain space, and when removed by evaporation or otherwise, would leave that space empty. The perfect adjustment, therefore, of the shape of one set of crystallizing bodies, to the shape of another set, as in the Portsoy granite, and their consolidation into one mass, is as strong a proof as could be desired, that they crystallized from a state of simple fluidity, such as, of all known causes, heat alone is able to produce.

82. This conclusion, however, does not rest on a single class of facts. It has been observed in many instances, that where granite and stratified rocks, such as primary schistus, are in contact, the latter are penetrated by veins of the former, which traverse them in various directions. These veins are of different dimensions, some being of the breadth of several yards, others of a few inches, or even tenths of an inch; they diminish as they recede from the main bo-

dy of the granite, to which they are always firmly united, conſtituting, indeed, a part of the ſame continued rock.

Theſe phenomena, which were firſt diſtinctly obſerved by Dr Hutton, are of great importance in geology, and afford a clear ſolution of the two chief queſtions concerning the relation between granite and ſchiſtus. As every vein muſt be of a date poſterior to the body in which it is contained, it follows, that the ſchiſtus was not ſuper-impoſed on the granite, after the formation of this laſt. If it be argued, that theſe veins, though poſterior to the ſchiſti, are alſo poſterior to the granite, and were formed by the infiltration of water in which the granite was diſſolved or ſuſpended; it may be replied, 1*mo*, That the power of water to diſſolve granite, is a poſtulatum of the ſame kind that we have ſo often, and for ſuch good reaſon, refuſed to concede; and, 2*do*, That in many inſtances the veins proceed from the main body of the granite *upwards* into the ſchiſtus; ſo that they are in planes much elevated in reſpect of the horizon, and have a direction quite oppoſite to that which the hypotheſis of infiltration requires. It remains certain, therefore, that the whole maſs of granite, and the veins proceeding from it, are coeval, and both of later formation than the ſtrata.

Now,

Now, this being eſtabliſhed, and the fluidity of the veins, when they penetrated into the ſchiſtus, being obvious, it neceſſarily follows, that the whole granite maſs was alſo fluid at the ſame time. But this can have been brought about only by ſubterraneous heat, which alſo impelled the melted matter againſt the ſuperincumbent ſtrata, with ſuch force as to raiſe them from their place, and to give them that highly inclined poſition in which they are ſtill ſupported by the granite, after its fluidity has ceaſed. Thus a concluſion, rendered probable by the cryſtallization of granite, is eſtabliſhed beyond all contradiction by the phenomena of granitic veins *.

83. With the granite, we ſhall conſider the proof of the igneous origin of all mineral ſubſtances as completed. Theſe ſubſtances, therefore, whether ſtratified or unſtratified, owe their conſolidation to the ſame cauſe, though acting with different degrees of energy. The ſtratified have been in general only ſoftened or penetrated by melted matter, whereas the unſtratified have been reduced into perfect fuſion.

84. In this general concluſion we may diſtinguiſh two parts, which, in their degree of certainty, differ perhaps ſomewhat from one another. The firſt of theſe, and that which ſtands higheſt in point of evidence, conſiſts of two propoſitions;

* Note xv.

propoſitions; namely, that the fluidity which preceded the conſolidation of mineral ſubſtances was SIMPLE, that is, it did not ariſe from the combination of theſe ſubſtances with any ſolvent; and, next, that after conſolidation, theſe bodies have been raiſed up by an expanſive force acting from below, and have by that means been brought into their preſent ſituation. Theſe two propoſitions ſeem to me to be ſupported by all the evidence that is neceſſary to conſtitute the moſt perfect demonſtration.

85. The other part of the general concluſion, that fire, or more properly heat, was the cauſe of the fluidity of theſe mineral bodies, and alſo of their ſubſequent elevation, is not perhaps to be conſidered as a truth ſo fully demonſtrated as the two preceding propoſitions; it is, no doubt, a matter of THEORY; or a portion of one of thoſe inviſible chains by which men ſeek to connect in the mind the ſtate of nature that is preſent, with the ſtates of it that are paſt; and participates of that uncertainty from which our reaſonings concerning ſuch cauſes as are not direct objects of perception, are hardly ever exempted. That it participates of this uncertainty in a very ſlight degree, will, however, be admitted, when it is conſidered that the cauſe aſſigned has been proved ſufficient for the effect; that the ſame is not true of any other known cauſe; and that

that this theory accounts, with ſingular ſimplicity and preciſion, for a ſyſtem of facts ſo various and complex, as that which is preſented by the natural hiſtory of the globe.

86. Neither can it be ſaid that the exiſtence of ſubterraneous heat is a principle aſſumed without any evidence, but that of the geological facts which it is intended to explain: on the contrary, it is proved by phenomena within the circle of ordinary experience, namely, thoſe of hot-ſprings, volcanoes, and earthquakes. Theſe leave no doubt of the exiſtence of heat, and of a moving and expanſive power, in the bowels of the earth; ſo that the only queſtions are, at what depth is this power lodged? to what extent, and with what intenſity, does it act? That it is lodged at a very conſiderable depth, is rendered probable by the permanency of ſome of the preceding phenomena: from the earlieſt times many fountains have retained their heat to the preſent day; and volcanoes, though they become extinguiſhed at length, have a very long period allotted for their duration. The cauſe of earthquakes is certainly a force that reſides very deep under the ſurface, otherwiſe the extent of the concuſſion could not be ſuch as has been obſerved in many inſtances.

87. The intenſity of volcanic fire, is another circumſtance that favours the opinion of its being

ing ſeated deep under the ſurface. That this intenſity is conſiderable, is certain from the experiments made by Sir James Hall on the fuſibility of whin-ſtone and lava; from which it appears, that the loweſt temperature in which either of theſe ſtones melt, is about 30° of Wedgewood's pyrometer. Some mineralogiſts have indeed affirmed, that lava is melted, not by the intenſity of the heat applied to it, but in conſequence of a certain combination formed between it and bituminous ſubſtances, in a manner which they do not attempt to explain, and which has indeed no analogy to any thing that is known. That a hypotheſis, formed in ſuch direct oppoſition to the moſt obvious principles of inductive reaſoning, ſhould have been imagined by a philoſopher who had examined the phenomena of Ætna and Veſuvius with much attention, and deſcribed them with great accuracy and truth, is more wonderful than that it ſhould have been adopted by mineralogiſts, whoſe views of nature may have been confined within a cabinet or a laboratory. It is, however, a hypotheſis, which, having never had any ſupport but from other hypotheſes, hardly merited the direct refutation that it has received from the experiments juſt mentioned.

88. But, if the intenſity of volcanic heat be ſuch as is here ſtated, it will be found very difficult

difficult to account for a fire of ſuch activity, and of ſuch long continuance in the ſame ſpot, by any decompoſition of mineral ſubſtances near the ſurface. In the place where this combuſtion is ſuppoſed to exiſt, it muſt be remembered, that there is no freſh ſupply of materials to replace thoſe that have been conſumed, and that, therefore, the original accumulation of theſe materials in one ſpot, muſt have been very unlike any thing that has ever been obſerved concerning the diſpoſition of minerals in the bowels of the earth.

89. If, on the other hand, we aſcribe the phenomena of volcanoes to the central heat, the account that may be given of them is ſimple, and conſiſtent with itſelf. According to all the appearances from which the exiſtence of ſuch heat has been inferred above, it is of a nature ſo far different from ordinary fire, that it may require no circulation of air, and no ſupply of combuſtible materials to ſupport it. It is not accompanied with inflammation or combuſtion, the great preſſure preventing any ſeparation of parts in the ſubſtances on which it acts, and the abſence of that elaſtic fluid without which heat ſeems to have no power to decompoſe bodies, even the moſt combuſtible, contributing to the unalterable nature of all the ſubſtances in the mineral regions. There, of conſequence, the only

only effects of heat are fusion and expansion; and that which forms the nucleus of the globe may therefore be a fluid mass, melted, but unchanged by the action of heat.

90. If, from the confines of this nucleus, we conceive certain fissures and openings to traverse the solid crust, and to issue at the surface of the earth, the vapours ascending through these may in time heat the sides of the tubes through which they pass to a vast distance from the lower extremities. It is, indeed, difficult to fix the limit to which this distance may extend, on account of the great difference between the rate at which heat moves when it has a fluid for its vehicle, and when it is left to make its way alone through a solid body. In the present case, the supply of heat is rapid, as being made by a vapour ascending through a tube of solid rock; and the dissipation of it slow, as arising from its transmission through the rock. The waste of heat is therefore small, compared with the supply, and grows smaller at every given point, the longer the stream of heated vapour has continued to flow. Such a stream, therefore, though it may at first be condensed within a small distance of its source, will in time reach higher and higher, and may at last be able to carry its heat to an immense distance from the place of its original derivation. Thus, it is easy to conceive,

ceive, that vapours from the mineral regions may convey their heat to reſervoirs of water near the ſurface of the earth, and may in that manner produce hot ſprings, and even boiling fountains, like thoſe of Rycum and Geyſer.

91. When, inſtead of a heated vapour, melted matter is thrown up through the *ſhafts* or *tubes*, which thus communicate with the mineral regions, veins of whinſtone and baſaltes are formed in the interior of the earth. When the melted matter reaches to the ſurface, it is thrown out in the form of lava, and all the other phenomena of volcanoes are produced.

Laſtly, Where melted matter of this kind, or vapours without being condenſed, have their progreſs obſtructed, thoſe dreadful concuſſions are produced, which ſeem to threaten the exiſtence even of the earth itſelf. Though terrible, therefore, to the preſent inhabitants of the globe, the earthquake has its place in the great ſyſtem of geological operations, and is part of a ſeries of events, eſſential, as will more clearly appear hereafter, to the general order, and to the preſervation of the whole.

Such, according to this theory, are the changes which have befallen mineral ſubſtances in the bowels of the earth; and though different for the ſtratified and unſtratified parts of thoſe

ſubſtances,

ſubſtances, they are connected together by the ſame *principle*, or explained by the ſame *cauſe*. It remains to conſider that part of the hiſtory of both which deſcribes their changes after their elevation to the ſurface; and here we ſhall find new cauſes introduced, which are more directly the ſubjects of obſervation, than thoſe hitherto treated of; cauſes, alſo, which act on all foſſils alike, and alike prepare them for their ultimate deſtination.

SEC-

SECTION III.

OF THE PHENOMENA COMMON TO STRATIFIED AND UNSTRATIFIED BODIES.

92. THE ſeries of changes which foſſil bodies are deſtined to undergo, does not ceaſe with their elevation above the level of the ſea; it aſſumes, however, a new direction, and from the moment that they are raiſed up to the ſurface, is conſtantly exerted in reducing them again under the dominion of the ocean. The ſolidity is now deſtroyed which was acquired in the bowels of the earth; and as the bottom of the ſea is the great laboratory, where looſe materials are mineralized and formed into ſtone, the atmoſphere is the region where ſtones are decompoſed, and again reſolved into earth.

This decompoſition of all mineral ſubſtances, expoſed to the air, is continual, and is brought about by a multitude of agents, both chemical and mechanical, of which ſome are known to us, and many, no doubt, remain to be diſcovered. Among the various aëriform fluid which compoſe our atmoſphere, one is already diſtinguiſhed as the grand principle of mineral decompoſition; the others are not inactive, and to them we muſt add moiſture, heat, and perhaps light; ſubſtances which, from their affinities to the elements of mineral bodies, have a power of entering into combination with them, and of thus diminiſhing the forces by which they are united to one another. By the action of air and moiſture, the metallic particles, particularly the iron, which enters in ſuch abundance into the compoſition of almoſt all foſſils, becomes oxydated in ſuch a degree as to loſe its tenacity; ſo that the texture of the ſurface is deſtroyed, and a part of the body reſolved into earth.

93. Some earths, again, ſuch as the calcareous, are immediately diſſolved by water; and though the quantity ſo diſſolved be extremely ſmall, the operation, by being continually renewed, produces a ſlow but perpetual corroſion, by which the greateſt rocks muſt in time be ſubdued. The action of water in deſtroying hard bodies into which it has obtained entrance, is much aſſiſted by the viciſſitudes of heat and cold, eſpecially when the latter extends as far as the point of congelation; for the water, when frozen, occupies a greater ſpace than before, and if the body is compact enough to refuſe room for this expanſion, its parts are torn aſunder by a repulſive force acting in every direction.

94. Beſides theſe cauſes of mineral decompoſition, the action of which we can in ſome meaſure

ſure trace, there are others known to us only by their effects.

We ſee, for inſtance, the pureſt rock cryſtal affected by expoſure to the weather, its luſtre tarniſhed, and the poliſh of its ſurface impaired, but we know nothing of the power by which theſe operations are performed. Thus alſo, in the precautions which the mineralogiſt takes to preſerve the freſh fracture of his ſpecimens, we have a proof how indiſcriminately all the productions of the foſſil kingdom are expoſed to the attacks of their unknown enemies, and we perceive how difficult it is to delay the beginnings of a proceſs which no power whatever can finally counteract.

95. The mechanical forces employed in the diſintegration of mineral ſubſtances, are more eaſily marked than the chemical. Here again water appears as the moſt active enemy of hard and ſolid bodies; and, in every ſtate, from tranſparent vapour to ſolid ice, from the ſmalleſt rill to the greateſt river, it attacks whatever has emerged above the level of the ſea, and labours inceſſantly to reſtore it to the deep. The parts looſened and diſengaged by the chemical agents, are carried down by the rains, and, in their deſcent, rub and grind the ſuperficies of other bodies. Thus water, though incapable of acting on hard ſubſtances by direct attrition, is the

caufe of their being fo acted on; and, when it defcends in torrents, carrying with it fand, gravel, and fragments of rock, it may be truly faid to turn the forces of the mineral kingdom againft itfelf. Every feparation which it makes is neceffarily permanent, and the parts once detached can never be united, fave at the bottom of the ocean.

96. But it would far exceed the limits of this fketch, to purfue the caufes of mineral decompofition through all their forms. It is fufficient to remark, that the confequence of fo many minute, but indefatigable agents, all working together, and having *gravity* in their favour, is a fyftem of univerfal decay and degradation, which may be traced over the whole furface of the land, from the mountain top to the fea fhore. That we may perceive the full evidence of this truth, one of the moft important in the natural hiftory of the globe, we will begin our furvey from the latter of thefe ftations, and retire gradually toward the former.

97. If the coaft is bold and rocky, it fpeaks a language eafy to be interpreted. Its broken and abrupt contour, the deep gulphs and falient promontories by which it is indented, and the proportion which thefe irregularities bear to the force of the waves, combined with the inequality of hardnefs in the rocks, prove, that the prefent

line

line of the ſhore has been determined by the action of the ſea. The naked and precipitous cliffs which overhang the deep, the rocks hollowed, perforated, as they are farther advanced in the ſea, and at laſt inſulated, lead to the ſame concluſion, and mark very clearly ſo many different ſtages of decay. It is true, we do not ſee the ſucceſſive ſteps of this progreſs exemplified in the ſtates of the ſame individual rock, but we ſee them clearly in different individuals; and the conviction thus produced, when the phenomena are ſufficiently multiplied and varied, is as irreſiſtible, as if we ſaw the changes actually effected in the moment of obſervation.

On ſuch ſhores, the fragments of rock once detached, become inſtruments of further deſtruction, and make a part of the powerful artillery with which the ocean aſſails the bulwarks of the land: they are impelled againſt the rocks, from which they break off other fragments, and the whole are thus ground againſt one another; whatever be their hardneſs, they are reduced to gravel, the ſmooth ſurface and round figure of which, are the moſt certain proofs of a *detritus* which nothing can reſiſt.

98. Again, where the ſea-coaſt is flat, we have abundant evidence of the degradation of the land in the beaches of ſand and ſmall gravel; the ſand banks and ſhoals that are continually

changing; the alluvial land at the mouths of the rivers; the bars that ſeem to oppoſe their diſcharge into the ſea, and the ſhallowneſs of the ſea itſelf. On ſuch coaſts, the land uſually ſeems to gain upon the ſea, whereas, on ſhores of a bolder aſpect, it is the ſea that generally appears to gain upon the land. What the land acquires in extent, however, it loſes in elevation; and, whether its ſurface increaſe or diminiſh, the depredations made on it are in both caſes evinced with equal certainty.

99. If we proceed in our ſurvey from the ſhores, inland, we meet at every ſtep with the fulleſt evidence of the ſame truths, and particularly in the nature and economy of rivers. Every river appears to conſiſt of a main trunk, fed from a variety of branches, each running in a valley proportioned to its ſize, and all of them together forming a ſyſtem of vallies, communicating with one another, and having ſuch a nice adjuſtment of their declivities, that none of them join the principal valley, either on too high or too low a level; a circumſtance which would be infinitely improbable, if each of theſe vallies were not the work of the ſtream that flows in it.

If indeed a river conſiſted of a ſingle ſtream, without branches, running in a ſtraight valley, it might be ſuppoſed that ſome great concuſſion,

cuſſion, or ſome powerful torrent, had opened at once the channel by which its waters are conducted to the ocean; but, when the uſual form of a river is conſidered, the trunk divided into many branches, which riſe at a great diſtance from one another, and theſe again ſubdivided into an infinity of ſmaller ramifications, it becomes ſtrongly impreſſed upon the mind, that all theſe channels have been cut by the waters themſelves; that they have been ſlowly dug out by the waſhing and eroſion of the land; and that it is by the repeated touches of the ſame inſtrument, that this curious aſſemblage of lines has been engraved ſo deeply on the ſurface of the globe.

100. The changes which have taken place in the courſes of rivers, are alſo to be traced, in many inſtances, by ſucceſſive platforms of flat alluvial land, riſing one above another, and marking the different levels on which the river has run at different periods of time. Of theſe, the number to be diſtinguiſhed, in ſome inſtances, is not leſs than four, or even five; and this neceſſarily carries us back, like all the operations we are now treating of, to an antiquity extremely remote: for, if it be conſidered, that each change which the river makes in its bed, obliterates at leaſt a part of the monuments of former changes, we ſhall be convinced, that

 only

only a ſmall part of the progreſſion can leave any diſtinct memorial behind it, and that there is no reaſon to think, that, in the part which we ſee, the beginning is included *.

101. In the ſame manner, when a river undermines its banks, it often diſcovers depoſites of ſand and gravel, that have been made when it ran on a higher level than it does at preſent. In other inſtances, the ſame ſtrata are ſeen on both the banks, though the bed of the river is now ſunk deep between them, and perhaps holds as winding a courſe through the ſolid rock, as if it flowed along the ſurface; a proof that it muſt have begun to ſink its bed, when it ran through ſuch looſe materials as oppoſed but a very inconſiderable reſiſtance to its ſtream. A river, of which the courſe is both ſerpentine and deeply excavated in the rock, is among the phenomena, by which the ſlow waſte of the land, and alſo the cauſe of that waſte, are moſt directly pointed out.

102. It is, however, where rivers iſſue through narrow defiles among mountains, that the identity of the ſtrata on both ſides is moſt eaſily recogniſed, and remarked at the ſame time with the greateſt wonder. On obſerving the Patowmack, where it penetrates the ridge of the Allegany mountains, or the Irtiſh, as it iſſues from the defiles of Altai, there is no man, however

ever

* Note xvi.

ever little addicted to geological ſpeculations, who does not immediately acknowledge, that the mountain was once continued quite acroſs the ſpace in which the river now flows; and, if he ventures to reaſon concerning the cauſe of ſo wonderful a change, he aſcribes it to ſome great convulſion of nature, which has torn the mountain aſunder, and opened a paſſage for the waters. It is only the philoſopher, who has deeply meditated on the effects which action long continued is able to produce, and on the ſimplicity of the means which nature employs in all her operations, who ſees in this nothing but the gradual working of a ſtream, that once flowed as high as the top of the ridge which it now ſo deeply interſects, and has cut its courſe through the rock, in the ſame way, and almoſt with the ſame inſtrument, by which the lapidary divides a block of marble or granite.

103. It is highly intereſting to trace up, in this manner, the action of cauſes with which we are familiar, to the production of effects, which at firſt ſeem to require the introduction of unknown and extraordinary powers; and it is no leſs intereſting to obſerve, how ſkilfully nature has balanced the action of all the minute cauſes of waſte, and rendered them conducive to the general good. Of this we have a moſt remarkable inſtance, in the proviſion made for preſerving the ſoil, or the coat of vegetable

getable mould, ſpread out over the ſurface of the earth. This coat, as it conſiſts of looſe materials, is eaſily waſhed away by the rains, and is continually carried down by the rivers into the ſea. This effect is viſible to every one; the earth is removed not only in the form of ſand and gravel, but its finer particles ſuſpended in the waters, tinge thoſe of ſome rivers continually, and thoſe of all occaſionally, that is, when they are flooded or ſwollen with rains. The quantity of earth thus carried down, varies according to circumſtances; it has been computed, in ſome inſtances, that the water of a river in a flood, contains earthy matter ſuſpended in it, amounting to more than the two hundred and fiftieth part of its own bulk *. The ſoil, therefore, is continually diminiſhed, its parts being tranſported from higher to lower levels, and finally delivered into the ſea. But it is a fact, that the ſoil, notwithſtanding, remains the ſame in quantity, or at leaſt nearly the ſame, and muſt have done ſo, ever ſince the earth was the receptacle of animal or vegetable life. The ſoil, therefore, is augmented from other cauſes, juſt as much, at an average, as it is diminiſhed by that now mentioned; and this augmentation evidently can proceed from nothing

* See Lehman, Traités de Phyſ. &c. tom. iii. p. 359. Note.

thing but the conftant and flow difintegration of the rocks. In the permanence, therefore, of a coat of vegetable mould on the furface of the earth, we have a demonftrative proof of the continual deftruction of the rocks; and cannot but admire the fkill, with which the powers of the many chemical and mechanical agents employed in this complicated work, are fo adjufted, as to make the fupply and the wafte of the foil exactly equal to one another.

104. Before we take leave of the rivers and the plains, we muft remark another fact, often obferved in the natural hiftory of the latter, and clearly evincing the former exiftence of immenfe bodies of ftrata, in fituations from which they have now entirely difappeared. The fact here alluded to is, the great quantity of round and hard gravel, often to be met with in the foil, under fuch circumftances, as prove, that it can only have come from the decompofition of rocks, that once occupied the very ground over which this gravel is now fpread. In the chalk country, for inftance, about London, the quantity of flints in the foil is every where great; and, in particular fituations, nothing but flinty gravel is found to a confiderable depth. Now, the fource from which thefe flints are derived is quite evident, for they are precifely the fame with thofe contained in the chalk beds, wherever

ever theſe laſt are found undiſturbed, and from the deſtruction of ſuch beds they have no doubt originated. Hence a great thickneſs of chalk muſt have been decompoſed, to yield the quantity of flints now in the ſoil of theſe countries; for the flints are but thinly ſcattered through the native chalk, compared with their abundance in the looſe earth. To afford, for example, ſuch a body of flinty gravel as is found about Kenſington, what an enormous quantity of chalk rock muſt have been deſtroyed?

105. This argument, which Dr Hutton has applied particularly to the chalk countries, may be extended to many others. The great plain of Crau, near the mouth of the Rhone, is well known, and was regarded with wonder, even in ages when the natural hiſtory of the globe was not an object of much attention. The immenſe quantity of large round gravel-ſtones, with which this extenſive plain is entirely covered, has been ſuppoſed, by ſome mineralogiſts, to have been brought down by the Durance, and other torrents, from the Alps; but, on further examination, has been found to be of the ſame kind that is contained in certain horizontal layers of pudding-ſtone, which are the baſis of the whole plain. It cannot be doubted, therefore, that the vaſt body of gravel ſpread over it, has originated from the deſtruction of

layers

layers of the ſame rock, which may perhaps have riſen to a great height above what is now the ſurface. Indeed, from knowing the depth of the gravel that covers the plain, and the average quantity of the like gravel contained in a given thickneſs of rock, one might eſtimate how much of the latter has been actually worn away. Whether data preciſe enough could be found, to give any weight to ſuch a computation, muſt be left for future inquiry to determine *.

106. In theſe inſtances, chalk and pudding-ſtone, by containing in them parts infinitely leſs deſtructible than their general maſs, have, after they are worn away, left behind them very unequivocal marks of their exiſtence. The ſame has happened in the caſe of mineral veins, where the ſubſtances leaſt ſubject to diſſolution have remained, and are ſcattered at a great diſtance from their native place. Thus gold, the leaſt liable to decompoſition of all the metals, is very generally diffuſed through the earth, and is found, in a greater or leſs abundance, in the ſand of almoſt all rivers. But the native place of this mineral is the ſolid rock, or the veins and cavities contained in the rock, and from thence it muſt have made its way into the ſoil. This, therefore, is another proof of the vaſt extent to which the degradation of the land, and of the

* Note xvii.

the rock, which is the bafis of it, has been carried; and confequently, of the great difference between the elevation and fhape of the earth's furface in the prefent, and in former ages.

107. The veins of tin furnifh an argument of the fame kind. The ores of this metal are very indeftructible, and little fubject to decompofition, fo that they remain very long in the ground without change. Where there are tin veins, as in Cornwall, the tin-ftone or tin-ore is found in great abundance in fuch vallies and ftreams as have the fame direction with the veins; and hence the *ftreaming*, as it is called, or wafhing of the earth, to obtain the tin-ftone from it. Now, if it be confidered, that none of this ore can have come into the foil but from parts of a vein actually deftroyed, it muft appear evident that a great wafte of thefe veins has taken place, and confequently of the fchiftus or granite in which they are contained.

108. Thefe leffons, which the geologift is taught in flat and open countries, become more ftriking, by the ftudy of thofe Alpine tracts, where the furface of the earth attains its greateft elevation. If we fuppofe him placed for the firft time in the midft of fuch a fcene, as foon as he has recovered from the impreffion made by the novelty and magnificence of the fpectacle before him, he begins to difcover the footfteps

footſteps of time, and to perceive, that the works of nature, uſually deemed the moſt permanent, are thoſe on which the characters of viciſſitude are moſt deeply imprinted. He ſees himſelf in the midſt of a vaſt ruin, where the precipices which riſe on all ſides with ſuch boldneſs and aſperity, the ſharp peaks of the granite mountains, and the huge fragments that ſurround their baſes, do but mark ſo many epochs in the progreſs of decay, and point out the energy of thoſe deſtructive cauſes, which even the magnitude and ſolidity of ſuch great bodies have been unable to reſiſt.

109. The reſult of a more minute inveſtigation, is in perfect uniſon with this general impreſſion. Whence is it, that the elevation of mountains is ſo obviouſly connected with the hardneſs and indeſtructibility of the rocks which compoſe them? Why is it, that a lofty mountain of ſoft and ſecondary rock is no where to be found; and that ſuch chains, as the Pyrenees or the Alps, never conſiſt of any but the hardeſt ſtone, of granite for inſtance, or of thoſe primarary ſtrata, which, if we are to credit the preceding theory, have been twice heated in the fires, and twice tempered in the waters, of the mineral regions? Is it not plain that this ariſes, not from any direct connection between the hardneſs of ſtones, and their height in the atmoſphere,

mosphere, but from this, that the waste and *detritus* to which all things are subject, will not allow soft and weak substances to remain long in an exposed and elevated situation? Were it not for this, the secondary rocks, being in position superincumbent on the primary, ought to be the highest of the two, and should cover the primary, (as they no doubt have at one time done), in the highest as well as the lowest situations, or among the mountains as well as in the plains.

110. Again, wherefore is it, that among all mountains, remarkable for their ruggedness and asperity, the rock, on examination, is always found of very unequal destructibility, some parts yielding to the weather, and to the other causes of disintegration, much more slowly than the rest, and having strength sufficient to support themselves, when left alone, in slender pyramids, bold projections, and overhanging cliffs? Where, on the other hand, the rock wastes uniformly, the mountains are similar to one another; their swells and slopes are gentle, and they are bounded by a waving and continuous surface. The intermediate degrees of resistance which the rocks oppose to the causes of destruction, produce intermediate forms. It is this which gives to the mountains, of every different species of rock,

a

a different habit and expreſſion, and which, in particular, has imparted to thoſe of granite that venerable and majeſtic character, by which they rarely fail to be diſtinguiſhed.

111. The ſtructure of the vallies among mountains, ſhews clearly to what cauſe their exiſtence is to be aſcribed. Here we have firſt a large valley, communicating directly with the plain, and winding between high ridges of mountains, while the river in the bottom of it deſcends over a ſurface, remarkable, in ſuch a ſcene, for its uniform declivity. Into this, open a multitude of tranſverſe or ſecondary vallies, interſecting the ridges on either ſide of the former, each bringing a contribution to the main ſtream, proportioned to its magnitude; and, except where a cataract now and then intervenes, all having that nice adjuſtment in their levels, (99.) which is the more wonderful, the greater the irregularity of the ſurface. Theſe ſecondary vallies have others of a ſmaller ſize opening into them; and, among mountains of the firſt order, where all is laid out on the greateſt ſcale, theſe ramifications are continued to a fourth, and even a fifth, each diminiſhing in ſize as it increaſes in elevation, and as its ſupply of water is leſs. Through them all, this law is in general obſerved, that where a higher valley joins a lower one, of the two angles which

it makes with the latter, that which is obtufe is always on the defcending fide; a law that is the fame with that which regulates the confluence of ftreams running on a furface nearly of uniform inclination. This alone is a proof that the vallies are the work of the ftreams; and indeed what elfe but the water itfelf, working its way through obftacles of unequal refiftance, could have opened or kept up a communication between the inequalities of an irregular and alpine furface.

112. Many more arguments, all leading to the fame conclufion, may be deduced from the general facts, known in the natural hiftory of mountains; and, if the Oreologift would trace back the progrefs of wafte, till he come in fight of that original ftructure, of which the remains are ftill fo vaft, he perceives an immenfe mafs of folid rock, naked and unfhapely, as it firft emerged from the deep, and incomparably greater than all that is now before him. The operation of rains and torrents, modified by the hardnefs and tenacity of the rock, has worked the whole into its prefent form; has hollowed out the vallies, and gradually detached the mountains from the general mafs, cutting down their fides into fteep precipices at one place, and fmoothing them into gentle declivities at another. From this has refulted a tranfportation of materials, which, both

for

for the quantity of the whole, and the magnitude of the individual fragments, muſt ſeem incredible to every one, who has not learned to calculate the effects of continued action, and to reflect, that length of time can convert accidental into ſteady cauſes. Hence fragments of rock, from the central chain, are found to have travelled into diſtant vallies, even where many inferior ridges intervene: hence the granite of Mount Blanc is ſeen in the plains of Lombardy, or on the ſides of Jura; and the ruins of the Carpathian mountains lie ſcattered over the ſhores of the Baltic *.

113. Thus, with Dr Hutton, we ſhall be diſpoſed to conſider thoſe great chains of mountains, which traverſe the ſurface of the globe, as cut out of maſſes vaſtly greater, and more lofty than any thing that now remains. The preſent appearances afford no data for calculating the original magnitude of theſe maſſes, or the height to which they may have been elevated. The neareſt eſtimate we can form is, where a chain or group of mountains, like thoſe of Roſa in the Alps, is horizontally ſtratified, and where, of conſequence, the undiſturbed poſition of the mineral beds enables us to refer the whole of the preſent inequalities of the ſurface to the operation of waſte or decay. Theſe

 mountains,

* Note xviii.

mountains, as they now ſtand, may not inaptly be compared to the pillars of earth which workmen leave behind them, to afford a meaſure of the whole quantity of earth which they have removed. As the pillars, (conſidering the mountains as ſuch), are in this caſe of leſs height than they originally were, ſo the meaſure furniſhed by them is but a limit, which the quantity ſought muſt neceſſarily exceed.

114. Such, according to Dr Hutton's theory, are the changes which the daily operations of waſte have produced on the ſurface of the globe. Theſe operations, inconſiderable if taken ſeparately, become great, by conſpiring all to the ſame end, never counteracting one another, but proceeding, through a period of indefinite extent, continually in the ſame direction. Thus every thing deſcends, nothing returns upward; the hard and ſolid bodies every where diſſolve, and the looſe and ſoft no where conſolidate. The powers which tend to preſerve, and thoſe which tend to change the condition of the earth's ſurface, are never *in equilibrio*; the latter are, in all caſes, the moſt powerful, and, in reſpect of the former, are like *living* in compariſon of *dead* forces. Hence the law of decay is one which ſuffers no exception: The elements of all bodies were once looſe and unconnected, and to the

the ſame ſtate nature has appointed that they ſhould all return.

115. It affords no preſumption againſt the reality of this progreſs, that, in reſpect of man, it is too ſlow to be immediately perceived: The utmoſt portion of it to which our experience can extend, is evaneſcent, in compariſon with the whole, and muſt be regarded as the momentary increment of a vaſt progreſſion, circumſcribed by no other limits than the duration of the world. TIME performs the office of *integrating* the infiniteſimal parts of which this progreſſion is made up; it collects them into one ſum, and produces from them an amount greater than any that can be aſſigned.

116. While on the ſurface of the earth ſo much is every where going to decay, no new production of mineral ſubſtances is found in any region acceſſible to man. The inſtances of what are called petrifactions, or the formation of ſtony ſubſtances by means of water, which we ſometimes obſerve, whether they be ferruginous concretions, or calcareous, or, as happens in ſome rare caſes, ſiliceous ſtalactites, are too few in number, and too inconſiderable in extent, to be deemed material exceptions to this general rule. The bodies thus generated, alſo, are no ſooner formed, than they become ſubject to waſte and diſſolution, like all the other hard ſubſtances in

 nature;

nature; ſo that they but retard for a while the progreſs by which they are all reſolved into duſt, and ſooner or later committed to the boſom of the deep.

117. We are not, however, to imagine, that there is no where any means of repairing this waſte; for, on comparing the concluſion at which we are now arrived, viz. that the preſent continents are all going to decay, and their materials deſcending into the ocean, with the propoſition firſt laid down, that theſe ſame continents are compoſed of materials which muſt have been collected from the decay of former rocks, it is impoſſible not to recogniſe two correſponding ſteps of the ſame progreſs; of a progreſs, by which mineral ſubſtances are ſubjected to the ſame ſeries of changes, and alternately waſted away and renovated. In the ſame manner, as the preſent mineral ſubſtances derive their origin from ſubſtances ſimilar to themſelves; ſo, from the land now going to decay, the ſand and gravel forming on the ſea-ſhore, or in the beds of rivers; from the ſhells and corals which in ſuch enormous quantities are every day accumulated in the boſom of the ſea; from the drift wood, and the multitude of vegetable and animal remains continually depoſited in the ocean: from all theſe we cannot doubt, that ſtrata are now forming in thoſe regions, to which

which nature ſeems to have confined the powers of mineral reproduction; from which, after being conſolidated, they are again deſtined to emerge, and to exhibit a ſeries of changes ſimilar to the paſt *.

118. How often theſe viciſſitudes of decay and renovation have been repeated, is not for us to determine: they conſtitute a ſeries, of which, as the author of this theory has remarked, we neither ſee the beginning nor the end; a circumſtance that accords well with what is known concerning other parts of the economy of the world. In the continuation of the different ſpecies of animals and vegetables that inhabit the earth, we diſcern neither a beginning nor an end; and, in the planetary motions, where geometry has carried the eye ſo far both into the future and the paſt, we diſcover no mark, either of the commencement or the termination of the preſent order †. It is unreaſonable, indeed, to ſuppoſe, that ſuch marks ſhould any where exiſt. The Author of nature has not given laws to the univerſe, which, like the inſtitutions of men, carry in themſelves the elements of their own deſtruction. He has not permitted, in his works, any ſymptom of infancy or of old age, or any ſign by which we may eſtimate either their future or their paſt duration. He may put an end, as he no doubt gave a begin-

 ning,

* Note xix. † Note xx.

ning, to the prefent fyftem, at fome determinate period; but we may fafely conclude, that this great *cataftrophe* will not be brought about by any of the laws now exifting, and that it is not indicated by any thing which we perceive.

119. To affert, therefore, that, in the economy of the world, we fee no mark, either of a beginning or an end, is very different from affirming, that the world had no beginning, and will have no end. The firft is a conclufion juftified by common fenfe, as well as found philofophy; while the fecond is a prefumptuous and unwarrantable affertion, for which no reafon from experience or analogy can ever be affigned. Dr Hutton might, therefore, juftly complain of the uncandid criticifm, which, by fubftituting the one of thefe affertions for the other, endeavoured to load his theory with the reproach of atheifm and impiety. Mr KIRWAN, in bringing forward this harfh and ill-founded cenfure, was neither animated by the fpirit, nor guided by the maxims of true philofophy. By the fpirit of philofophy, he muft have been induced to reflect, that fuch poifoned weapons as he was preparing to ufe, are hardly ever allowable in fcientific conteft, as having a lefs direct tendency to overthrow the fyftem, than to hurt the perfon of an adverfary, and to wound, perhaps incurably, his mind, his reputation, or his peace.

By

By the maxims of philosophy, he must have been reminded, that, in no part of the history of nature, has any mark been discovered, either of the beginning or the end of the present *order;* and that the geologist sadly mistakes, both the object of his science and the limits of his understanding, who thinks it his business to explain the means employed by INFINITE WISDOM for establishing the laws, which now govern the world.

By attending to these obvious considerations, Mr Kirwan would have avoided a very illiberal and ungenerous proceeding; and, however he might have differed from Dr Hutton as to the *truth* of his opinions, he would not have censured their *tendency* with such rash and unjustifiable severity.

But, if this author may be blamed for wanting the temper, or neglecting the rules, of philosophic investigation, he is hardly less culpable, for having so slightly considered the scope and spirit of a work which he condemned so freely. In that work, instead of finding the world represented as the result of necessity or chance, which might be looked for, if the accusations of atheism or impiety were well founded, we see every where the utmost attention to discover, and the utmost disposition to admire, the instances of wise and beneficent design,

ſign manifeſted in the ſtructure, or economy of the world. The enlarged views of theſe, which his geological ſyſtem afforded, appeared to Dr Hutton himſelf as its moſt valuable reſult. They were the parts of it which he contemplated with greateſt delight; and he would have been leſs flattered, by being told of the ingenuity and originality of his theory, than of the addition which it had made to our knowledge of *final cauſes*. It was natural, therefore, that he ſhould be hurt by an attempt to accuſe him of opinions, ſo different from thoſe which he had always taught; and if he anſwered Mr Kirwan's attack with warmth or aſperity, we muſt aſcribe it to the indignation excited by unmerited reproach.

120. But to return to the natural hiſtory of the earth: Though there be in it no *data*, from which the commencement of the preſent order can be aſcertained, there are many by which the exiſtence of that order may be traced back to an antiquity extremely remote. The beds of primitive ſchiſtus, for inſtance, contain ſand, gravel, and other materials, collected, as already ſhewn, from the diſſolution of mineral bodies; which bodies, therefore, muſt have exiſted long before the oldeſt part of the preſent land was formed. Again, in this gravel we ſometimes find pieces of ſandſtone, and of other compound rocks, by which we are of courſe carried back a ſtep farther, ſo as to reach

a

a ſyſtem of things, from which the preſent is the third in ſucceſſion; and this may be conſidered as the moſt ancient epocha, of which any memorial exiſts in the records of the foſſil kingdom.

121. Next in the order of time to the conſolidation of the primary ſtrata, we muſt place their elevation, when, from being horizontal, and at the bottom of the ſea, they were broken, ſet on edge, and raiſed to the ſurface. It is even probable, as formerly obſerved, that to this ſucceeded a depreſſion of the ſame ſtrata, and a ſecond elevation, ſo that they have twice viſited the ſuperior, and twice the inferior regions. During the ſecond immerſion, were formed, firſt, the great bodies of pudding-ſtone, that in ſo many inſtances lie immediately above them; and next were depoſited the ſtrata that are ſtrictly denominated ſecondary.

122. The third great event, was the raiſing up of this compound body of old and new ſtrata from the bottom of the ſea, and forming it into the dry land, or the continents, as they now exiſt *. Contemporary with this, we muſt ſuppoſe the injection of melted matter among the ſtrata, and the conſequent formation of the cryſtallized and unſtratified rocks, namely, the granite, metallic veins, and veins of porphyry and whinſtone.

* Note xxi.

whinſtone. This, however, is to be conſidered as embracing a period of great duration; and it muſt always be recollected, that veins are found of very different formation; ſo that when we ſpeak generally, it is perhaps impoſſible to ſtate any thing more preciſe concerning their antiquity, than that they are poſterior to the ſtrata, and that the veins of whinſtone ſeem to be the moſt recent of all, as they traverſe every other.

123. In the fourth place, with reſpect to time, we muſt claſs the facts that regard the detritus and waſte of the land, and muſt carefully diſtinguiſh them from the more ancient phenomena of the mineral kingdom. Here we are to reckon the ſhaping of all the preſent inequalities of the ſurface; the formation of hills of gravel, and of what have been called tertiary ſtrata, conſiſting of looſe and unconſolidated materials; alſo collections of ſhells not mineralized, like thoſe in Turaine; ſuch petrifactions as thoſe contained in the rock of Gibraltar, on the coaſt of Dalmatia, and in the caves of Bayreuth. The bones of land animals found in the ſoil, ſuch as thoſe of Siberia, or North America, are probably more recent than any of the former *.

124. Theſe phenomena, then, are all ſo many marks of the lapſe of time, among which the principles of geology enable us to diſtinguiſh a certain

* Note xxii.

certain order, fo that we know fome of them to be more, and others to be lefs diftant, but without being able to afcertain, with any exactnefs, the proportion of the immenfe intervals which feparate them. Thefe intervals admit of no comparifon with the aftronomical meafures of time; they cannot be expreffed by the revolutions of the fun or of the moon; nor is there any fynchronifm between the moft recent epochas of the mineral kingdom, and the moft ancient of our ordinary chronology.

125. On what is now faid is grounded another objection to Dr Hutton's theory, namely, that the high antiquity afcribed by it to the earth, is inconfiftent with that fyftem of chronology which refts on the authority of the Sacred Writings. This objection would no doubt be of weight, if the high antiquity in queftion were not reftricted merely to the globe of the earth, but were alfo extended to the human race. That the origin of mankind does not go back beyond fix or feven thoufand years, is a pofition fo involved in the narrative of the Mofaic books, that any thing inconfiftent with it, would no doubt ftand in oppofition to the teftimony of thofe ancient records. On this fubject, however, geology is filent; and the hiftory of arts and fciences, when traced as high as any authentic monuments extend, refers the

the beginnings of civilization to a date not very different from that which has juſt been mentioned, and infinitely within the limits of the moſt recent of the epochas, marked by the phyſical revolutions of the globe.

On the other hand, the authority of the Sacred Books ſeems to be but little intereſted in what regards the mere antiquity of the earth itſelf; nor does it appear that their language is to be underſtood literally concerning the *age* of that body, any more than concerning its *figure* or its *motion*. The theory of Dr Hutton ſtands here preciſely on the ſame footing with the ſyſtem of COPERNICUS; for there is no reaſon to ſuppoſe, that it was the purpoſe of revelation to furniſh a ſtandard of geological, any more than of aſtronomical ſcience. It is admitted, on all hands, that the Scriptures are not intended to reſolve phyſical queſtions, or to explain matters in no way related to the morality of human actions; and if, in conſequence of this principle, a conſiderable latitude of interpretation were not allowed, we ſhould continue at this moment to believe, that the earth is flat; that the ſun moves round the earth; and that the circumference of a circle is no more than three times its diameter.

It is but reaſonable, therefore, that we ſhould extend to the geologiſt the ſame liberty of ſpeculation,

culation, which the aſtronomer and mathematician are already in poſſeſſion of; and this may be done, by ſuppoſing that the chronology of MOSES relates only to the human race. This liberty is not more neceſſary to Dr Hutton than to other theoriſts. No ingenuity has been able to reconcile the natural hiſtory of the globe with the opinion of its recent origin; and accordingly the coſmologies of Kirwan and De Luc, though contrived with more mineralogical ſkill, are not leſs forced and unſatisfactory than thoſe of Burnet and Whiſton.

126. It is impoſſible to look back on the ſyſtem which we have thus endeavoured to illuſtrate, without being ſtruck with the novelty and beauty of the views which it ſets before us. The very plan and ſcope of it diſtinguiſh it from all other theories of the earth, and point it out as a work of great and original invention. The ſole object of ſuch theories has hitherto been, to explain the manner in which the preſent laws of the mineral kingdom were firſt eſtabliſhed, or began to exiſt, without treating of the manner in which they now proceed, and by which their continuance is provided for. The authors of theſe theories have accordingly gone back to a ſtate of things altogether unlike the preſent, and have confined their reaſonings, or their

their fictions, to a crisis which never has existed but once, and which never can return. Dr Hutton, on the other hand, has guided his investigation by the philosophical maxim, *Causam naturalem et assiduam quærimus, non raram et fortuitam.* His theory, accordingly, presents us with a system of wise and provident economy, where the same instruments are continually employed, and where the decay and renovation of fossils being carried on at the same time in the different regions allotted to them, preserve in the earth the conditions essential for the support of animal and vegetable life. We have been long accustomed to admire that beautiful contrivance in nature, by which the water of the ocean, drawn up in vapour by the atmosphere, imparts, in its descent, fertility to the earth, and becomes the great cause of vegetation and of life; but now we find, that this vapour not only fertilizes, but creates the soil; prepares it from the solid rock, and, after employing it in the great operations of the surface, carries it back into the regions where all its mineral characters are renewed. Thus, the circulation of moisture through the air, is a prime mover, not only in the annual succession of the seasons, but in the great geological cycle, by which the waste and reproduction of entire continents is circumscribed. Perhaps a more striking view than this, of the wisdom

dom that presides over nature, was never presented by any philosophical system, nor a greater addition ever made to our knowledge of final causes. It is an addition which gives consistency to the rest, by proving, that equal foresight is exerted in providing for the whole and for the parts, and that no less care is taken to maintain the constitution of the earth, than to preserve the tribes of animals and vegetables which dwell on its surface. In a word, it is the peculiar excellence of this theory, that it ascribes to the phenomena of geology an order similar to that which exists in the provinces of nature with which we are best acquainted; that it produces seas and continents, not by accident, but by the operation of regular and uniform causes; that it makes the decay of one part subservient to the restoration of another, and gives stability to the whole, not by perpetuating individuals, but by reproducing them in succession.

127. Again, in the detail of this theory, and the ample induction on which it is founded, we meet with many facts and observations, either entirely new, or hitherto very imperfectly understood. Thus, the veins which proceed from masses of granite, and penetrate the incumbent schistus, had either escaped the observation of former mineralogists, or the importance of the phenomenon had been entirely overlooked. Dr

Hutton has defcribed the appearances with great accuracy, and drawn from them the moft interefting conclufions. At the junction of the primary and fecondary ftrata, the facts which he has noted had been obferved by others; but no one I think had fo fully underftood the language which they fpeak, or had fo clearly perceived the confequences that neceffarily follow from them. He is the firft who diftinctly pointed out the characters which diftinguifh whinftone from lava, and who explained the true relation that fubfifts between thefe fubftances. He alfo difcovered the induration of the ftrata, in contact with veins of whin, and the charring of the coal in their vicinity. His theory alfo enabled him to determine the affinity of whinftone and granite to one another, and their relation to the other great bodies of the mineral kingdom.

To the obfervations of the fame excellent geologift, we are indebted for the knowledge of the general and important fact, that all the hard fubftances of the mineral kingdom, when elevated into the atmofphere, have a tendency to decay and are fubject to a difintegration and wafte, to which no limit can be fet but that of their entire deftruction; that no provifion is made on the furface for repairing this wafte, and that there, no new foffil is produced; that the formation of all the varied fcenery which the

furface

ſurface of the earth exhibits, depends on the operation of cauſes, the momentary exertions of which are familiar to us, though we knew not before the effects which their accumulated action was able to produce. Theſe are facts in the natural hiſtory of the earth, the diſcovery of which is due to Dr Hutton; and, ſhould we lay all further ſpeculation aſide, and conſider the theory of the earth as a work too great to be attempted by man, we muſt ſtill regard the phenomena and laws juſt mentioned, as forming a ſolid and valuable addition to our knowledge.

128. If we would compare this theory with others, as to the inviſible agents which it employs, we muſt conſider, that fire and water are the two powers which all of them muſt make uſe of, ſo that they can differ from one another only by the way in which they combine theſe powers. In Dr Hutton's ſyſtem, water is firſt employed to depoſite and arrange, and then fire to conſolidate, mineralize, and laſtly, to elevate the ſtrata; but, with reſpect to the unſtratified or cryſtallized ſubſtances, the action of fire only is recogniſed. The ſyſtem having leaſt affinity to this is the Neptunian, which aſcribes the formation of all minerals to the action of water alone, and extends this hypotheſis even to the unſtratified rocks. Here, therefore, the action of fire is entirely excluded; and the Neptuniſts

 have

have certainly made a great ſacrifice to the love of truth, or of paradox, in rejecting the aſſiſtance of ſo powerful an auxiliary *.

129. In the ſyſtems which employ the agency of the latter element, we are to look for a greater reſemblance to that of Dr Hutton, though many and great marks of diſtinction are eaſily perceived. In the coſmologies, for example, of Leibnitz and Buffon, fire and water are both employed, as well as in this; but they are employed in a reverſe order. Theſe philoſophers introduce the action of fire firſt, and then the action of water, which is to invert the order of nature altogether, as the conſolidation of the rocks muſt be poſterior to their ſtratification. Indeed, the theory of Buffon is ſingularly defective: beſides inverting the order of the two great operations of ſtratification and conſolidation, and of courſe giving no real explanation of the latter, it gives no account of the elevation, or highly inclined poſition of the ſtrata; it makes no diſtinction between ſtratified and unſtratified bodies, nor does it offer any but the moſt unſatisfactory explanation of the inequalities earth's ſurface. This ſyſtem, therefore, a very diſtant reſemblance to the theory †.

130. The ſyſtem of Lazzaro Moro has been remarked as approaching nearer to this theory

* Note xxiii. † Note xxiv.

ry than any other; and it is certain, that one very important principle is common to them both. The theory of the Italian geologiſt was chiefly directed to the explanation of the remains of marine animals, which are found in mountains far from the ſea; and it appears to have been ſuggeſted to him by the phenomena of the *Campi Phlegræi*, and by the production of the new iſland of *Santorini* in the Archipelago. He accordingly ſuppoſes, that the iſlands and continents have been all raiſed up, like the above-mentioned iſland, from the bottom of the ſea, by the force of volcanic fire: that theſe fires began to burn under the bottom of the ocean, ſoon after the creation of the world, when as yet the ocean covered the whole earth: that they at firſt elevated a portion of the land; and in this primitive land no ſhells are found, as the original ocean was deſtitute of fiſh. The volcanoes continuing to burn, under the ſea, after the creation of animated nature, the ſtrata that were then raiſed up by their action were full of ſhells and other marine objects; and, from the violence with which they were elevated, aroſe the contortions and inclined poſition which they frequently poſſeſs *.

 This

* Dé Croſtacei, et degli altri Marini Corpi, che ſi trovano ſu' Monti: di Ant. Lazzaro Moro. Vinezia, 1740.

This ſyſtem is imperfect, as it makes no peculiar proviſion for the conſolidation of the ſtrata, which, according to it, as well as the Neptunian ſyſtem, muſt be aſcribed to the action, not of fire, but of water. No account is given of the mineralization of the ſhells found in the ſtrata, or of the difference between them and the ſhells found looſe at the bottom of the ſea; and no diſtinction is made between ſtratified and unſtratified ſubſtances. But, with all this, Lazzaro Moro has certainly the merit of having perceived, that ſome other power than that which depoſited the ſtrata, muſt have been employed for their elevation, and that they have endured the action of a diſturbing force.

131. From this compariſon it appears, that Dr Hutton's theory is ſufficiently diſtinct, even from the theories which approach to it moſt nearly, to merit, in the ſtricteſt ſenſe, the appellation of *new* and *original.* There are indeed few inventions or diſcoveries, recorded in the hiſtory of ſcience, to which nearer approaches were not made before they were fully unfolded. It therefore very well deſerves to be diſtinguiſhed by a particular name; and, if it behoves us to follow the analogy obſerved in the names of the two great ſyſtems, which at preſent divide the opinions of geologiſts, we may join Mr Kirwan in calling this the PLUTONIC SYSTEM. For

my

my own part, I would rather have it characteriz-ed by a lefs fplendid, but jufter name, that of the HUTTONIAN THEORY.

132. The circumftance, however, which gives to this theory its peculiar character, and exalts it infinitely above all others, is the introduction of the principle of preffure, to modify the effects of heat when applied at the bottom of the fea. This is in fact the key to the grand enigma of the mineral kingdom, where, while one fet of phenomena indicates the action of fire, another fet, equally remarkable, feems to exclude the poffibility of that action, by prefenting us with mineral fubftances, in fuch a ftate as they could never have been brought into by the operation of the fires we fee at the furface of the earth. Thefe two claffes of phenomena are reconciled together, by admitting the power of compreffion to confine the volatile parts of bodies when heat is applied to them, and to force them, in many inftances, to undergo fufion, inftead of being calcined or diffipated by burning or inflammation. In this hypothefis, which fome affect to confider as a principle gratuitoufly affumed, there appears to me nothing but a very fair and legitimate generalization of the properties of heat. Combuftion and inflammation are chemical proceffes, to which other conditions are required, befides the prefence of a high temperature. The

ftate of the mineral regions makes it reafonable to prefume, that thefe conditions are wanting in the bowels of the earth, where, of confequence, we have a right to look for nothing but expanfion and fufion, the only operations which feem effential to heat, and infeparable from the application of it, in certain degrees, to certain fubftances. Though this principle, therefore, had no countenance from analogy, the admirable fimplicity, and the unity, which it introduces into the phenomena of geology, would fufficiently juftify the application of it to the theory of the earth.

As another excellence of this theory, I may, perhaps, be allowed to remark, that it extends its confequences beyond thofe to which the author of it has himfelf adverted, and that it affords, which no geological theory has yet done, a fatisfactory explanation of the fpheroidal figure of the earth *.

133. Yet, with all thefe circumftances of originality, grandeur, and fimplicity in its favour, with the addition of evidence as demonftrative as the nature of the fubject will admit, this theory has probably many obftacles to overcome, before it meet the general approbation. The greatnefs of the objects which it fets before us, alarms the imagination; the powers which it fuppoies to be lodged in the fubterraneous regions,

* Note xxv.

gions; a heat which has ſubdued the moſt refractory rocks, and has melted beds of marble and quartz; an expanſive force, which has folded up, or broken the ſtrata, and raiſed whole continents from the bottom of the ſea; theſe are things with which, however certainly they may be proved, the mind cannot ſoon be familiariſed. The change and movement alſo, which this theory aſcribes to all that the ſenſes declare to be moſt unalterable, raiſe up againſt it the ſame prejudices which formerly oppoſed the belief in the true ſyſtem of the world; and it affords a curious proof, how little ſuch prejudices are ſubject to vary, that as ARISTARCHUS, an ancient follower of that ſyſtem, was charged with impiety for moving the everlaſting VESTA from her place, ſo Dr Hutton, nearly on the ſame ground, has been ſubjected to the very ſame accuſation. Even the length of time which this theory regards as neceſſary to the revolutions of the globe, is looked on as belonging to the marvellous; and man, who finds himſelf conſtrained by the want of time, or of ſpace in almoſt all his undertakings, forgets, that in theſe, if in any thing, the riches of nature reject all limitation *.

The evidence which muſt be oppoſed to all theſe cauſes of incredulity, cannot be fully underſtood without much ſtudy and attention. It

* NOTE XXVI.

It requires not only a careful examination of particular inftances, but comprehenfive views of the whole phenomena of geology; the comparifon of things very remote with one another; the interpretation of the *obfcure* by the *luminous*, and of the *doubtful* by the *decifive* appearances. The geologift muft not content himfelf with examining the infulated fpecimens of his cabinet, or with purfuing the nice fubtleties of mineralogical arrangement; he muft ftudy the relations of foffils, as they actually exift; he muft follow nature into her wildeft and moft inacceffible abodes; and muft felect, for the places of his obfervations, thofe points, from which the variety and gradation of her works can be moft extenfively and accurately explored. Without fuch an exact and comprehenfive furvey, his mind will hardly be prepared to relifh the true theory of the earth. "*Naturæ enim vis atque majeftas omnibus momentis fide caret, fi quis modo partes atque non totam complectatur animo* *."

134. If indeed this theory of the earth is as well founded as we fuppofe it to be, the lapfe of time muft neceffarily remove all objections to it, and the progrefs of fcience will only develope its evidence more fully. As it ftands at prefent,

* PLIN. Hift. Nat. lib. vii. cap. i.

ſent, though true, it muſt be ſtill imperfect; and it cannot be doubted, that the great principles of it, though eſtabliſhed on an immoveable baſis, muſt yet undergo many modifications, requiring to be limited, in one place, or to be extended, in another. A work of ſuch variety and extent cannot be carried to perfection by the efforts of an individual. Ages may be required to fill up the bold outline which Dr Hutton has traced with ſo maſterly a hand; to detach the parts more completely from the general maſs; to adjuſt the ſize and poſition of the ſubordinate members; and to give to the whole piece the exact proportion and true colouring of nature.

This, however, in length of time, may be expected from the advancement of ſcience, and from the mutual aſſiſtance which parts of knowledge, ſeemingly the moſt remote, often afford to one another. Not only may the obſervations of the mineralogiſt, in tracts yet unexplored, complete the enumeration of geological facts; and the experiments of the chemiſt, on ſubſtances not yet ſubjected to his analyſis, afford a more intimate acquaintance with the nature of foſſils, and a meaſure of the power of thoſe chemical agents to which this theory aſcribes ſuch vaſt effects: but alſo, from other ſciences, leſs directly connected with the natural hiſtory

of

of the earth, much information may be received. The accurate geographical maps and ſurveys which are now making; the ſoundings; the obſervations of currents; the barometrical meaſurements, may all combine to aſcertain the reality, and to fix the quantity of thoſe changes which terreſtrial bodies continually undergo. Every new improvement in ſcience affords the means of delineating more accurately the face of nature as it *now* exiſts, and of tranſmitting, to future ages, an account, which may be compared with the face of nature as it ſhall *then* exiſt. If, therefore, the ſcience of the preſent times is deſtined to ſurvive the phyſical revolutions of the globe, the HUTTONIAN THEORY may be confirmed by hiſtorical record; and the author of it will be remembered among the illuſtrious few, whoſe ſyſtems have been verified by the obſervations of ſucceeding ages, ſupported by facts unknown to themſelves, and eſtabliſhed by the deciſions of a tribunal, ſlow, but infallible, in diſtinguiſhing between truth and falſehood.

NOTES AND ADDITIONS.

NOTES AND ADDITIONS.

NOTE I. § 2.

Origin of calcareous rocks.

134. IT has been afferted, that Dr Hutton went further than is ftated at § 2., and maintained all calcareous matter to be *originally* of animal formation. This pofition, however, is fo far from being laid down by Dr Hutton, that it belongs to an inquiry which he carefully avoided to enter on, as being altogether beyond the limits of philofophical inveftigation.

He has indeed no where treated of the *firft origin* of any of the earths, or of any fubftance whatfoever, but only of the transformations which bodies have undergone fince the prefent laws of nature were eftablifhed. He confidered this laft as all that a fcience, built on experiment and obfervation, can poffibly extend to; and willingly left, to more prefumptuous inquirers, the tafk of carrying their reafonings beyond the boundaries of nature, and of unfolding the properties of the chaotic fluid, with as much minutenefs of detail, as if they were deſcribing the circumſtances of a chemical proceſs which they had actually witneſſed.

The idea of calcareous matter which really belongs to the Huttonian Theory, is, that in all the changes which the terraqueous globe has undergone in paſt ages, this matter exiſted, as it does now, either in the form of limeſtone and marble, or in the compoſition of other ſtones, or in the ſtate of corals, ſhells, and bones of animals. It may be true, that there is no particle of calcareous matter, at preſent exiſting on the ſurface of the earth, that has not, at ſome time, made a part of an animal body; but of this we can have no certainty, nor is it of any importance that we ſhould. It is enough to know, that the rocks of marble and limeſtone contain in general marks of having been formed from materials collected at the bottom of the ſea; and of this a ſingle cockle-ſhell, or piece of coral, found included in a rock, is a ſufficient proof with reſpect to the whole maſs of which it makes a part..

The principal object which Dr Hutton had in view when he ſpoke of the maſſes of marble and limeſtone, as compoſed of the calcareous matter of marine bodies *, was to prove, that they had been all formed at the bottom of the

* Theory of the Earth, vol. i. p. 23, 24.

the ſea, and from materials there depoſited. His general concluſion is, " that all the ſtrata of the earth, not only thoſe conſiſting of ſuch calcareous maſſes, but others ſuperincumbent upon theſe, have had their origin at the bottom of the ſea, by the collection of ſand and gravel, of ſhells, of coralline and cruſtaceous bodies, and of earths and clays variouſly mixed, or ſeparated and accumulated. This is a general concluſion, well authenticated by the appearances of nature, and highly important in the natural hiſtory of the earth *."

135. In his Geological Eſſays, Mr Kirwan ſays, that " ſome geologiſts, as Buffon, and of late Dr Hutton, have excluded calcareous earth from the number of the primeval, aſſerting the maſſes of it we at preſent behold to proceed from ſhell-fiſh. But, in addition to the unfounded ſuppoſition, that ſhell-fiſh, or any animals, poſſeſs the power of producing any ſimple earth, theſe philoſophers ſhould have conſidered, that, before the exiſtence of any fiſh, the ſtony maſſes that incloſe the baſon of the ſea, muſt have exiſted; and, among theſe, there is none in which calcareous earth is not found. Dr Hutton endeavours to *evade* this argument, by ſuppoſing the world we now inhabit to have ariſen from the

K ruins

* Theory of the Earth, vol. i. p. 26.

ruins and fragments of an anterior, without pointing at any original. If we are thus to proceed *in infinitum*, I ſhall not pretend to follow him; but, if he ſtops any where, he will find the ſame argument equally to occur *."

The argument here employed would certainly be concluſive againſt any one, who, in diſputing about the *firſt origin* of things, ſhould deny that the calcareous is as ancient as any other of the ſimple earths. But this has nothing to do with Dr Hutton's ſpeculations, which, as has been juſt ſaid, never extended to the *firſt origin* of ſubſtances, but were confined entirely to their changes; ſo that what he aſſerts concerning the calcareous rocks, is no more than that thoſe which we now ſee have been formed from looſe materials, depoſited at the bottom of the ſea. It was not therefore in order to *evade* Mr Kirwan's argument, as the preceding paſſage would lead us to believe, that he ſuppoſed the world which we now inhabit to have ariſen from the ruin and waſte of an anterior world; but it was becauſe this ſeemed to him a concluſion which neceſſarily followed from the phenomena of geology, and it was a concluſion that he had deduced long before he heard of Mr Kirwan's objections to his ſyſtem. Inſtead of an *evaſion*, therefore,

* Geol. Eſſays, p. 13.

therefore, any one who confiders the fubject fairly, will fee, in Dr Hutton's reafoning, nothing but the caution of a philofopher, who wifely confines his theory within the fame limits by which nature has confined his experience and obfervation.

It is neverthelefs true, that Dr Hutton has fometimes expreffed himfelf as if he thought that the prefent calcareous rocks are all compofed of animal remains *. This conclufion, however, is more general than the facts warrant; and, from fome incorrectnefs or ambiguity of language, is certainly more general than he intended. The idea of calcareous rocks, on which he argues throughout his whole theory, is precifely that which is ftated in the preceding article.

NOTE II. § 6.

Origin of coal.

136. The vegetable origin of coal feems to be fufficiently proved by the reafoning in § 5. and 6.; and that reafoning will appear ftill more fatisfactory, from what is faid at § 28. and 29. concerning the confolidation of this foffil. Dr Hutton has treated both of the matter of coal

 and

* Theory of the Earth, vol. i. p. 23.

and of its consolidation, Part. I. Chap. 8. of his Theory of the Earth *.

The notion, however, that coal is of vegetable origin, is not peculiar to this theory, but has been for some time the prevailing opinion. Buffon supposes this mineral to be formed from vegetable and animal substances, the oil and fat of which have been converted into bitumen by the action of acids †. A fundamental mistake, however, is committed by this author, and by M. Gensanne, (author of the natural history of Languedoc), on whose observations he greatly relies, in considering coal as consisting of bitumen united to earth, thus omitting the only ingredient essential to coal, namely the carbon or charcoal. This may truly be considered as the essential part, because coal may exist without bitumen, as in the instance of blind-coal, but not without charcoal.

Another theory of coal, very analogous to Dr Hutton's, is that of Arduino, professor of mineralogy at Venice, in which he supposes it formed from vegetable and animal remains from the land and sea, but chiefly from the latter ‡. This

* Vol i. p. 558, &c.

† Hist. Nat. des Mineraux, tom. i. p. 429. 4to edit.

‡ Saggio Fisico-mineralogico del Sig. Giov. Arduino; Atti di Siena, tom. v. p. 228, 281, &c.

This theory of coal is contained in Dr Hutton's, in which the animal and vegetable remains muſt be ſuppoſed to come both from the earth and the ſea. It ſeems to be without any good reaſon that Arduino conſiders the ſea as the chief ſource of theſe materials. His remarks, however, are very ingenious, and deſerving of attention.

Theſe accounts of the origin of coal are all nearly the ſame; it is in what relates to the diſtinction between the common coal, in which there is no ligneous ſtructure, and thoſe varieties of it in which that ſtructure is apparent, and again in explaining the conſolidation of both, that the theory, laid down here, is peculiar.

137. Some other mineralogiſts refer one of the ingredients of coal to the vegetable kingdom, but not the other. Unable to reſiſt the conviction which ariſes from the fibrous ſtructure of parts of ſtrata, and even entire ſtrata of coal, they have ſuppoſed, that wood, which had been ſomehow buried in the earth, or perhaps depoſited at the bottom of the ſea, had become impregnated with bitumen, which laſt, however, they conſider as of mineral origin. This appears to be the opinion of Lehman, and alſo of ſome very late writers. There ſeems, however, to be hardly leſs reaſon for referring the origin of one part of coal to the vegetable or animal kingdom

 than

than another. The two laſt are certainly capable of furniſhing both the carbonic and bituminous parts; and therefore, to derive theſe from different ſources, is at leaſt a very unneceſſary complication of hypotheſes.

138. Another explanation of coal, very different from any of the preceding, has lately been advanced and ſet up in oppoſition to the Huttonian Theory. Mr Kirwan *, the only mineralogiſt, I believe, who has attempted to derive both the carbonic and bituminous matter of coal from the mineral kingdom, diſtinguiſhes between wood-coal and mineral-coal, and gives a theory entirely new of the formation of the latter. Wood-coal is that in which the ligneous ſtructure is ſo apparent, as to leave no doubt of its vegetable origin; mineral coal is that in which no ſuch ſtructure can be diſcovered, and is the ſame which Dr Hutton derives from the vegetable juices, and other remains, comminuted, diſperſed, carried into the ſea, and there precipitated, ſo as to unite with different proportions of earth, and to become afterwards mineralized.

Theſe two ſpecies of coal, which the Huttonian theory conſiders as gradations of the ſame ſubſtance, Mr Kirwan regards as perfectly diſtinct, conſtituting two minerals, of an origin and

* Geol. Eſſays, eſſay vii. p. 290.

and formation entirely different. He therefore endeavours to aſcertain the diſtinguiſhing characters of each, conſidered geologically.

139. But here the leading diſtinction, implied in all the reſt, that the two kinds of coal are never found in the ſame bed, but always in different ſituations, and with different laws of ſtratification, is expreſsly contradicted by matter of fact. Coal, as is ſaid above, with its ligneous texture quite apparent, and coal with no ſuch ſtructure viſible, are often found in the ſame ſeam, are brought up from the ſame mine, and united in the ſame ſpecimen. I have a ſpecimen from a bed of coal, in the Iſle of Sky, found under a baſaltic rock, conſiſting of a ligneous part, which graduates into one in which there is no veſtige of a fibrous texture, and in which the ſurface is ſmooth and gloſſy, with a fracture almoſt vitreous. The upper part of the ſpecimen is therefore perfect wood-coal, and the under part perfect mineral-coal, in the language of Mr Kirwan; at the ſame time that the tranſition from the one to the other is made by inſenſible degrees. This ſpecimen, were it perfectly ſolitary, is ſufficient to prove the identity of the two ſpecies of coal we are now ſpeaking of, and to ſhew, that the difference between them is accidental not eſſential. The ſpecimen, however, is far from being ſolitary; the number of ſimi-

lar appearances is ſo great, as hardly to have eſcaped the obſervation of any mineralogiſt. Mr Kirwan admits, that wood-coal is often found under baſaltes*; but what is eſſential to be remarked is, that, in this inſtance, we have both the wood-coal, and the common mineral-coal, lying under that rock, and the one paſſing gradually into the other. It appears, indeed, that many of the facts which Mr Kirwan produces, in treating of what he calls *carboniferous* ſoils, are quite inconſiſtent with the diſtinction he would make between wood-coal and mineral-coal †.

140. It is, however, true, that there are inſtances in which the wood-coal, or foſſil-wood, as it is uſually called, forms entire beds, quite unconnected with the ordinary coal, and ſtratified in ſome reſpects differently. Such is the Bovey coal in Devonſhire, the wood-coal in the north of Ireland, and perhaps the Surturbrandt of Iceland. With reſpect to the Bovey coal, it does by no means anſwer to one of Mr Kirwan's remarks viz. that late obſervations have aſcertained, that no ſuch parallеliſm of the beds, as in mineral-coal, nor even any diſtinct number of ſtrata, is found. In the Bovey coal, the number of ſtrata is very well defined, by beds of clay regularly interpoſed; but as to the extent of theſe beds,

* Geol. Eſſays, p. 310. † *ibid.* p 311.

beds, the coal having been worked only at one place, and by an open pit, without any extenſive ſubterraneous excavation, nothing is known with certainty.

In the Bovey coal too, I muſt obſerve, though its beds have the ligneous ſtructure very diſtinct, the clay interpoſed between theſe beds, which is but little indurated, contains a great deal of coaly matter, in the form of thin flakes, interſperſed through it. So far as I know, there are no mineral veins nor ſhifts, nor any bed of indurated ſtone, that accompany this coal; ſo that, though one cannot doubt of its vegetable origin, ſome doubt may be entertained concerning the nature of the mineralizing operations, to which it has been ſubjected. The conſideration of theſe, however, does not belong to the preſent argument; and the peculiarities of this ſemimineralized coal, as it may be called, have nothing to do with the general queſtion, whether wood-coal and mineral-coal are the ſame ſubſtance; about which queſtion, if the gradations are properly conſidered, I think, no reaſonable doubt can remain.

141. One of Mr Kirwan's objections to the vegetable origin of coal, is founded on this fact, that there is, in the muſeum at Florence, a cellular ſandſtone, the cells of which are filled with genuine mineral coal. "Could this, (adds he) have

have been originally wood * ?" The anſwer to the interrogatory propoſed here as a *reductio ad abſurdum*, is, that moſt undoubtedly it may have been wood. Sandſtone with charred wood, that is, with wood-coal in it, is not an uncommon phenomenon in coal countries. I have ſeen a ſpecimen of this kind from the Hales Quarry, near Edinburgh, conſiſting of a piece of charred wood, imbedded in ſandſtone; the wood was much altered, but the remains of its fibrous ſtructure were diſtinctly viſible. This affords a perfect commentary on the ſpecimen in the Florence cabinet.

142. If then it be granted, as I think it muſt, that the two kinds of coal we have been ſpeaking of are of the ſame origin, it is not very neceſſary to enter on a refutation of Mr Kirwan's theory with reſpect to either of them. His account of the formation of mineral-coal, however, is ſo ſingular, that it cannot be paſſed over without remark.

Mr Kirwan ſuppoſes, 1mo, That natural carbon was originally contained in many mountains of the granite and porphyritic order, and alſo in ſiliceous ſchiſtus; and might, by diſintegration and decompoſition, be ſeparated from the ſtony particles. 2do, That both petrol and carbon are often contained in trap, ſince hornblend, which

* Geol. Eſſays, p. 321.

which has lately been found to contain carbon, very frequently enters into its compofition.

"My opinion (adds he) is, that coal mines, or ftrata of coal, as well as the mountains in which they are found, owe their origin to the difintegration of primeval mountains, either now totally deftroyed, or whofe height and bulk, in confequence of fuch difintegration, are confiderably leffened; and that thefe rocks, anciently deftroyed, contained moft probably a far larger proportion of carbon and petrol than thofe of the fame denomination now contain, fince their difintegration took place at fo early a period *.

"By the decompofition of thefe mountains, the feltfpar and hornblend were converted into clay; the bituminous particles, thus fet free, reunited, and were abforbed, partly by the argil, but chiefly by the carbonaceous matter, with which they have the greateft affinity. The carbonic and bituminous particles, thus united, being difficultly mifcible with water, and fpecifically heavier, funk through the moift, pulpy, incoherent argillaceous maffes, and formed the loweft ftratum," &c.

Such is Mr Kirwan's theory of the formation of coal, and nobody I think will difpute the originality of it.

143. To

* Geol. Effays. p. 328, &c.

143. To enter on a formal refutation of an opinion ſo loaded with objections, would be a taſk as irkſome as unneceſſary. A few obſervations will ſuffice.

The notion of the great degradation of mountains, involved in this hypotheſis, is the part of it to which I am leaſt diſpoſed to object. But I cannot help reminding Mr Kirwan, that the effects of waſte are not ſuppoſed leſs in this, than in Dr Hutton's theory; and that he has aſſumed the very principle, of which that theory makes ſo much uſe, though he has reſerved to himſelf, as it ſhould ſeem, the right of denying it, when it does not accord with his ſyſtem. It is indeed worth while to compare what is ſaid concerning the degradation of mountains, in the above quotations, and ſtill more fully in the book itſelf, with what is advanced concerning their indeſtructibility, in another paſſage of the ſame volume *:

" All mountains are not ſubject to decay; for inſtance, ſcarce any of thoſe that conſiſt of red granite. The ſtone of which the Runic rocks are formed, have withſtood decompoſition for two thouſand years, as their characters evince," &c.

" Baſaltic pillars in general, bid defiance to decay," &c. He goes on to deny every ſtep of the degradation of land, by which it is waſted, carried

* Page 436.

carried into the ſea, and ſpread out over its bottom, though all theſe are neceſſary *poſtulata* in his theory of the formation of coal. One can be at no loſs about eſtimating the value of a ſyſtem, in which ſuch groſs inconſiſtencies make a neceſſary part.

144. The quantity of hornblend and ſiliceous ſchiſtus, neceſſary to be decompoſed, in order to produce the coal ſtrata preſently exiſting, is enormous, and would lead to an eſtimate of what is worn away from the primeval mountains, far exceeding any thing that Dr Hutton has ſuppoſed. It is true that Mr Kirwan, never at all embarraſſed about preſerving a ſimilitude between nature as ſhe is now, and as ſhe was heretofore, lays it down, that the part of the primeval mountains which is worn away, contained much more carbon than the part which is left behind. This, however, is an arbitrary ſuppoſition ; and ſince, in this ſyſtem, ſuch ſuppoſitions are ſo eaſily admitted, why may we not conceive, in the primeval mountains, a more copious ſource of carbonic matter than hornblend or ſiliceous ſchiſtus ? We have but to imagine, that the *diamond* exiſted among theſe mountains in ſuch abundance, as to conſtitute large rocks. This ſtone being made up of pure, or highly concentrated carbon, the adamantine ſummits of a ſingle ridge, by their decompoſition,

decompoſition, might afford a carbonic baſis, ſufficient for the coal beds of all the ſurrounding plains.

145. We may alſo object to Mr Kirwan, that the ſiliceous part of the mountains has not been chemically diſſolved; it has been only abraded and worn away. Mechanical action has reduced the quartz to gravel and ſand, but has not produced on it any chemical change. The carbon, therefore, could not be let looſe. Experiment, indeed, might be employed, to determine whether the ſiliceous matter of the ſecondary, and of the primary ſtrata contains this ſubſtance in the ſame proportion.

Again, a more fatal ſymptom can hardly be imagined in any theory, than that, when the circumſtances of the phenomena to be explained are *a little* changed, the theory is under the neceſſity of changing *a great deal.* Now, this is what happens to Mr Kirwan's theory, in the attempt made to explain by it the ſtratum of coal deſcribed in the *Annales de Chimie* *, as cutting a mountain of argillaceous ſtrata in two, at about three-fourths of its height. This ſtratum, Mr Kirwan ſays, muſt have been formed by *tranſudation* from the ſuperior part of the mountain †. Beſides that this is a gratuitous ſuppoſition of a thing,

* Tom. xi. p. 272.

† Geol. Eſſays, p. 338.

thing, without example, it involves in it an absurdity, which becomes evident the moment the queſtion is aſked, What occupied the place of the coal-bed before the tranſudation from the upper part of the mountain? Has the *liquid coal*, as it percolated through the upper ſtrata, expelled any ſubſtance from the place it now occupies? or has it been powerful enough to raiſe up, or to float, as it were, the upper part of the mountain?

The ſituation of this bed of coal is not ſingular, and its formation is eaſily explained on Dr Hutton's theory. It is part of a ſtratum of coal, which has been depoſited, like all others, at the bottom of the ſea; from whence certain cauſes, of very general operation, have raiſed it up, together with the attending ſtrata: theſe ſtrata have ſince been all cut down, and worn away by the operations of the ſurface; and the mountain, with the coal ſtratum in the middle of it, is a part of them which has been left behind. There is no wonder, that a coal ſtratum ſhould be found alternating with others, in a mountain, any more than in the bowels of the earth, and no more need of a ſeparate explanation *.

146. After

* This ſtratum of coal, which is deſcribed by HASSENFRATZ, is remarkable for being in a mountain which reſts immediately on primary ſchiſtus and granite.

146. After all, it may be aſked, for what purpoſe is it that ſo many incongruous and ill-ſupported hypotheſes are thus piled on one another? is it only to avoid aſcribing the carbonic and bituminous matter of coal to a ſubſtance in which we know with certainty that ſuch matter reſides in great abundance, in order to derive it from other ſubſtances, in which a ſubtle analyſis has ſhewn, that it exiſts in a very ſmall proportion? Such reaſoning is ſo great a treſpaſs on every principle of common ſenſe, not to ſay of ſound philoſophy, that, to beſtow any time on the refutation of it, is, in ſome degree, to fall under the ſame cenſure.

Note iii. § 7.

Primitive mountains.

147. The enumeration of the different kinds of primary ſchiſtus, at § 7., is not propoſed as at all complete. It will be leſs defective, however, if we add to it *talcoſe ſchiſtus*, and *lapis-ollaris* or *potſtone* *.

148. The rocks called here by the name of primary, were firſt diſtinguiſhed, as forming the baſis

* Kirwan's Mineralogy, vol. i. p. 155.

basis of all the great chains of mountains, and as constituting a separate division of the mineral kingdom, by J. G. LEHMAN, director of the Prussian mines. See his work, intituled, *Essai d'une Histoire Naturelle de Couches de la Terre* *. These rocks were regarded by Lehman as parts of the original *nucleus* of the globe, which had undergone no alteration, but remained now such as they were at first created; and, agreeably to this supposition, he bestowed on them, and on the mountains composed of them, the name of primitive. He remarks, nevertheless, their distribution into beds, either perpendicular to the horizon, or highly inclined, and the super-position of the secondary, and horizontal strata. However mineralogists may now differ in their theories from Lehman, they must consider this distinction as a great step in the science of geology, and very material to the right arrangement of the natural history of the earth.

149. Several mineralogists have agreed with him in the supposition, that these rocks are a part of the original structure of the globe, and prior to all organized matter. Of this number is PALLAS †; and also De Luc, who applies the term

 primordial

* Tom iii. p. 239, &c. The French translation is in 1759, but the original preface is dated at Berlin 1756.

† Observations sur la Formation des Montagnes.

primordial to the rocks in queſtion, and conſiders them as neither ſtratified nor formed by water *. In his ſubſequent writings, however, he admits their formation from aqueous depoſition, as the Neptuniſts do in general, but holds them to be more ancient than organized bodies.

150. PINI, profeſſor of natural hiſtory at Milan, has denied the ſtratification of primitive mountains, in a memoir on the mineralogy of St Gothard, and in another on the revolutions of the globe †. His reaſonings are oppoſed by SAUSSURE ‡, and are certainly, in many reſpects, very open to attack. They proceed on a comparifon between the diviſion of rocks, by what is called the planes of their ſtratification, and their diviſion by tranſverſe fiſſures; two things, which he thinks ſo much alike, that they ought not to be referred to different cauſes; and, as the one cannot be regarded as the effect of aqueous depoſition, ſo neither ſhould the other. This is a very fallacious argument, becauſe it confounds two things that are eſſentially different; and,

* Lettres Phyſ ſur l'Hiſtoire de la Terre, tom. ii. p. 06.

† Memoria ſulle Rivoluzione del Globo Terreſtre; Memorie della Societa Italiana, tom. v. p. 222, &c.

‡ Voyages aux Alpes, tom. IV. § 1881.

and, inftead of inquiring about a matter of fact, inquires about its caufe. The truth is, that the difpute has arifen from not diftinguifhing the granite from the fchiftus mountains, and from involving both under the name of primitive. M. Pini feems to be in the right, when he holds the granite of St Gothard to be unftratified ; but it is without any good reafon, that he would extend the fame conclufion to the fchiftus of that mountain. CHARPENTIER, and Sauffure, in his laft two volumes, contend even for the ftratification of granite *.

As the confent, if not univerfal, is very general for the ftratification of the primary fchiftus, and the fact itfelf abundantly obvious, in almoft all the inftances I have ever met with, I have not confidered it as neceffary to enter here into any argument on this fubject.

NOTE IV. § 8.

Primary ftrata not primitive.

151. An account of the facts referred to § 8., may be found in Hutton's Theory, vol. i.

 p. 332,

* See NOTE XV. on Granite.

p. 332, &c. To what is there ſaid, of the ſhells contained in the primary limeſtone of Cumberland, I muſt add, that I have ſince had an opportunity of verifying the conjecture, that the limeſtone rock, in which the ſhells were found, near the head of *Coniſton* Lake, is part of the ſame body of ſtrata, where ſhells were found, in a quarry between Ambleſide and Low-wood. The limeſtone of that quarry contains ſeveral marine objects; it is in ſtrata declining about 10° from the perpendicular, toward the S. E., and forms a belt, ſtretching acroſs the country from N. E. to S. W.

In a quarry where the argillaceous ſchiſtus, on the ſouth ſide of this limeſtone belt, is worked for pavement, are impreſſions of what I think may ſafely be accounted marine objects; they have the form of ſhells, are much indurated, and full of pyrites. They ſeem to be of the ſame kind with the impreſſions ſaid to be found in a ſlate quarry, near the village of Mat in Switzerland *.

Another ſpot, affording inſtances of ſhells in primary limeſtone, is in Devonſhire. On the ſea-ſhore on the eaſt ſide of Plymouth Dock, oppoſite to Stonehouſe, I found a ſpecimen of ſchiſtoſe micaceous limeſtone, containing a ſhell of

* Hutton's Theory, vol. i. p. 327.

of the bivalve kind: it was ftruck off from the folid rock, and cannot poffibly be confidered as an adventitious foffil.

Now, no rocks can be more decidedly primary than thofe about Plymouth. They confift of calcareous ftrata, in the form either of marble or micaceous limeftone, alternating with varieties of the fame fchiftus, which prevails through Cornwall to the weft, and extends eaftward into Dartmoor, and on the fea-coaft, as far as the Berry-head. Thefe all interfect the horizontal plane, in a line from eaft to weft nearly; they are very erect, thofe at Plymouth being elevated to the north.

Though, therefore, the remains of marine animals are not frequent among the primary rocks, they are not excluded from them; and hence the exiftence of fhell-fifh and zoophytes, is clearly proved to be anterior to the formation even of thofe parts of the prefent land which are juftly accounted the moft ancient.

152. The rocks which contain fand or gravel, or which are of a granulated texture, muft alfo be confidered as carrying in themfelves a teftimony of the moft unequivocal kind, of their being derived from the *detritus* and wafte of former rocks. Now, the fact ftated in the text, concerning fand found in fchiftus, moft juftly accounted primary, might be exemplified by actual

 reference

reference to many ſpots on the earth's ſurface. A few ſuch will be ſufficient in this place.

St Gothard is a central point, in one of the greateſt tracts of primary mountains on the face of the earth, yet arenaceous ſtrata are found in its vicinity. Between Ayrolo and the Hoſpice of St Gothard, Sauſſure found a rock, compoſed of an arenaceous or granular paſte, including in it hornblend and garnets. He is ſomewhat unwilling to give the name *gres* to this ſtone, which M. Beſſon had done; but he neverthe-leſs deſcribes it as having a granulated ſtructure *.

Among the moſt indurated rocks that compoſe the mountains of this iſland, many are arenaceous. Thus, on the weſtern coaſt of Scotland, the great body of high and rugged mountains on the ſhores of Araſaig, &c. from Ardnamurchan to Glenelg, conſiſts, in a great meaſure, of a granitic ſandſtone, in vertical beds. This ſtone ſometimes occupies great tracts; at other times it is alternated with the micaceous, or other varieties of primary ſchiſtus; it occurs, likewiſe, in ſeveral of the iſlands, and is a foſſil which we hardly find deſcribed or named by the writers on mineralogy. Much, alſo, of a

* Voyages aux Alpes, tom. iv. § 1822.

a highly indurated, but granulated quartz, is found in ſeveral places in Scotland, in beds or ſtrata, alternated with the common ſchiſtus of the mountains. Remarkable inſtances of this may be ſeen on the north ſide of the ferry of Balachuliſh, and again on the ſea-ſhore at Cullen. At the latter, the ſtrata are remarkably regular, alternating with different ſpecies of ſchiſtus. At the former, the quartz is ſo pure, that the ſtone has been miſtaken for marble.

Theſe examples are perhaps ſufficient; but I muſt add, that in the micaceous and talcoſe ſchiſti themſelves, thin layers of ſand are often found, interpoſed between the layers of mica or talc. I have a ſpecimen, from the ſummit of one of the higheſt of the Grampian mountains, where the thin plates, of a talcky or aſbeſtine ſubſtance, are ſeparated by layers of a very fine quartzy ſand, not much conſolidated.

The mountain from which it was brought, conſiſts of vertical ſtrata, much interſected by quartz veins. It is impoſſible to doubt, in this inſtance, that the thin plates of the one ſubſtance, and the ſmall grains of the other, were depoſited together at the bottom of the ſea, and that they were alike produced from the degradation of rocks, more ancient than any which now exiſt.

153. In the Neptunian ſyſtem, as improved by WERNER, an attempt is made to take off the force of ſuch inſtances as are produced in § 8, 9, and 151, &c. by diſtinguiſhing rocks, as to their formation, into three different orders, the primitive, the intermediate, and the ſecondary, or, to ſpeak more properly, into primary, ſecondary, and tertiary. The ſame mineralogiſt diſtinguiſhes, among the materials of theſe rocks, between what he terms chemical and mechanical depoſites. By mechanical depoſites, are underſtood ſand, gravel, and whatever bears the mark of fracture and attrition ; by chemical depoſites, thoſe which are regularly cryſtallized, or which have a tendency to cryſtallization, and in which the action of mechanical cauſes cannot be traced. This diſtinction is founded in nature, and proceeds on real and palpable differences ; but the application made of it to the three kinds of ſtrata juſt enumerated, ſeems by no means entitled to the ſame praiſe.

The primitive rocks contain, it is ſaid, none but chemical depoſites, and are entirely compoſed of them : the intermediate contain a mixture of both, and alſo ſome veſtiges of organized bodies : the ſecondary conſiſt almoſt entirely of the mechanical, or of the remains of ſuch bodies, with little of the chemical. The firſt of theſe, then, are held to contain no mark or veſtige

tige whatſoever of any thing more ancient than themſelves, and are, in the ſtricteſt ſenſe, primeval, or formed of the firſt materials, depoſited by the immenſe ocean which originally encompaſſed the globe.

After them were formed the intermediate, moſtly conſiſting of chemical depoſites, but containing alſo ſome animal remains, and ſome ſpoils from the land, ſubjected to the various kinds of deſtruction, which even then made a part of the order of nature. Theſe rocks, it is alleged, are chiefly argillaceous, are lets indurated than the primary, and not interſected by veins of quartz.

The ſecondary were formed from the remains of the other two, and contain more mechanical depoſites than any other.

This ſketch of what I underſtand to be Werner's opinion concerning the different formation of the ſtrata, is chiefly taken from a view of his ſyſtem, in the *Journal de Phyſique* for 1800.

154. The main objection to the diſtinction here made between the primary and the intermediate ſtrata, is founded on the facts that have been juſt ſtated. The ſandſtone of St Gothard is from a country having every character of a primary one in the higheſt perfection. The inſtances I have mentioned from the Highlands of Scotland, are from mountains, leſs elevated indeed

indeed than the Alps, but where the rock is micaceous, talcofe, or filiceous, in planes erect to the horizon, and interfected by veins of quartz. The fhells from Plymouth are from a rock, that Werner would, I think, admit to be truly primitive. Thofe from the lakes, alfo, are from the centre of a country, occupied by porphyry, fchorl, hornftone-fchiftus, and many others, about the order of which there can be no difpute. It is true, that in this tract there are argillaceous ftrata, of the kind that might be accounted intermediate, were they not interpofed among thofe that are certainly primary ; and this very intermixture fhews, how little foundation there is for the diftinction attempted to be made between the formation of the one and of the other. If there is any principle in mineralogy, which may be confidered as perfectly afcertained, it is, that rocks fimilarly ftratified, and alternated with one another, are of the fame formation.

Hence we conclude, that there is *no order of ftrata yet known*, that does not contain proofs of the exiftence of more ancient ftrata. We fee nothing, in the ftrict fenfe, primitive. It muft be underftood, that what is here faid has no reference to granite, which I do not confider as a ftratified rock, and in which neither the remains of organized bodies, nor fand, have I

believe

believe been ever found; though ſome inſtances will be hereafter mentioned, where granite contains fragments of other ſtones, viz. of different kinds of primary ſchiſtus.

To the inſtances of ſand involved in primary ſchiſtus, I might have added many from the rocks of that order on the coaſt of Berwickſhire, of which mention is ſo often made in theſe Illuſtrations; but I wiſhed to draw the evidence from thoſe rocks that are moſt unequivocally primary, and to which the Wernerian diſtinction of *intermediate* could not poſſibly be applied.

If any one aſſert, as M. de Luc has done, that ſand is a chemical depoſite, a certain mode of cryſtallization which quartz ſometimes aſſumes, let him draw the line which ſeparates ſand from gravel; and let him explain why quartz, in the form of ſand, is not found in mineral veins, in granite, nor in baſaltes, that is, in none of the ſituations where the appearances of cryſtallization are moſt general and beſt aſcertained.

Note v. § 10.

Tranſportation of the materials of the ſtrata.

155. The great tranſportation or *travelling* of the materials of the ſtrata, ſuppoſed by Dr Hutton,

ton, has been treated as abſurd by ſome of his opponents, particularly De Luc and Kirwan. Theſe philoſophers ſeem not to have obſerved, that their own ſyſtem, and indeed every ſyſtem which derives the ſecondary ſtrata from the primary, involves a tranſportation of materials, hardly leſs than is ſuppoſed in the Huttonian theory, and a degradation of the primeval mountains, in many inſtances much greater. To form ſome notion of this degradation, it muſt be recollected, that the primeval mountains, which furniſhed the materials of the ſecondary ſtrata in the plains, cannot have ſtood in the place now occupied by theſe plains. This is obvious; and therefore we muſt neceſſarily regard the ſecondary ſtrata as derived from the primitive mountains which are the neareſt to them, and of which a part ſtill remains. This part is ſufficient to define the baſe of the original mountains; and the quantity of the ſecondary ſtrata which ſurround them may help us to make ſome eſtimate of their height. Let us take, for inſtance, the extenſive tract of ſecondary country about Newcaſtle, where coal mines have been ſunk through a ſucceſſion of ſecondary ſtrata, to the depth of more than a thouſand feet. This ſecondary country may be conſidered as comprehending almoſt the whole of the counties of Northumberland and Durham, and probably as

extending

extending very far under the part of the German Ocean which washes their coasts; and the whole strata composing it must be derived, on the hypothesis we are now considering, from the Cheviot Hills, on one side, and from those in the high parts of Westmoreland and Cumberland on the other, comprehending the Alston-Moor Hills, and the large group of primary mountains, so well known from the sublime and romantic scenery of the *Lakes*. Now, the mountains which stood on this base, had not only to supply the materials for the tract already mentioned, on the east, but had also their contingent to furnish to the plains on the west and north; the Cheviots to Roxburghshire and Berwickshire; the Northumberland mountains to the coal strata about Whitehaven, and along the sea-coast to Lancashire. On the whole, we shall not exceed the truth, if we suppose, that the secondary strata, at the feet of the above mountains, are six or seven times more extensive than the base of the mountainous tract. If then we take the medium depth of these secondary strata to be one thousand feet, it is evident, that the mass of stone which composes them, if it were placed on the same base with the primitive mountains, would reach to the height of six thousand feet. This is supposing the mass to preserve the breadth of its base uniformly to the

the ſummit; but if it be ſuppoſed to taper, as mountains uſually do, we muſt multiply this ſix thouſand by three, in order to have the height of theſe primeval mountains, which, therefore, were originally elevated not leſs than eighteen thouſand feet: in height, therefore, they once rivalled the Cordelieras, and are now but poorly repreſented by the hills of Skidaw and Helvellyn. It were eaſy to ſhew, that this eſtimate is ſtill below the reſult that ſtrictly follows from the Neptunian hypotheſis; but it is unneceſſary to proceed further, than to prove, that the principle of the degradation of mountains, is involved in that hypotheſis to an exceſſive and improbable degree; and that the ſupporters of it, have either been guilty of the inconſiſtency of refuſing to Dr Hutton the moderate uſe of a principle, which they themſelves employ in its utmoſt extent, or of not having ſufficiently adverted to the conſequences of their own ſyſtem.

156. The formation of ſecondary ſtrata from the degradation of the contiguous mountains, on cloſe examination, is ſubject to many other difficulties of the ſame kind. Mountains of ſecondary ſtrata, and nearly horizontal, are found in this iſland of the height of three thouſand feet. Such are Ingleborough, Wharnſide, and perhaps ſome others on the weſt of Yorkſhire. The

whole

whole chain, indeed, for ſecondary mountains, is of great elevation. The ſtrata are of limeſtone, and of a very coarſe-grained ſandſtone, alternating with it. No mountains can more clearly point out, that the ſtrata of which they conſiſt were once continued quite acroſs the vallies which now ſeparate them; and hence, if the materials of thoſe ſtrata were indeed furniſhed from any contiguous primitive mountains, the latter muſt have been, out of all proportion, higher than any mountains now in Britain.

157. Thus, a great degradation of the primitive mountains, and of courſe a great travelling of their materials, is proved to make a neceſſary part of the Neptunian theory. The extent of this travelling or tranſportation may be rendered more evident, if we apply a ſimilar mode of reaſoning to larger portions of the globe. The north-weſt of Europe furniſhes us an inſtance of a very extenſive tract of ſecondary country, comprehending the greater part of Britain, the whole of Flanders and Holland, part of Germany, the northern provinces of France, and probably the bed of the German Ocean, at leaſt for a great extent. Within this circle almoſt all is ſecondary, and on the ſides of it all round are placed ridges or groups of primitive mountains, namely, the mountains of Auvergne,

at

at leaſt in part, and going round by the eaſt, the Alps, the Voſges, the Hartz, the Highlands and Weſtern Iſlands of Scotland, the hilly countries of Cumberland, Wales and Cornwall. This zone of primitive mountains, on the ſuppoſition of the Neptuniſts, muſt have riſen up in the form of iſlands in the great ocean, that originally covered the earth, forming a kind of circular Archipelago, including in its boſom a ſea, which was from ſeven to five hundred miles in diameter. Over the whole of this extent, the *detritus* of the above mountains muſt have been carried, in order to form the flat interjacent countries which are now expoſed to our view. Such then, even on their own ſuppoſition, is the extent to which the Neptuniſts muſt admit that the materials of the primeval mountains were tranſported by the ocean.

158. This tranſportation of materials, may not be ſo great as that which is involved in Dr Hutton's theory, but is ſuch as ſhould make the enemies of his ſyſtem conſider, how nearly the principles they *muſt* introduce, agree with thoſe that they *would* reject. This is one fact out of many, which ſhews, that there is at preſent a much nearer agreement between the ſyſtems of geology, than between their authors.

159. To

159. To theſe facts, demonſtrating the great tranſportation of foſſils in ſome former conditions of the globe, we may add another, recogniſed by all mineralogiſts. The animal *exuviæ* contained in limeſtone and marble, are often known to belong to ſeas, extremely remote from the countries where they are now found. In the chalk-beds of England, in the limeſtones of France, a great proportion of the petrifactions belong to the tropical ſeas, and appear to have been brought from the vicinity of the equator. Buffon obſerves, that of the foſſil ſhells found in France, it has been diſputed, whether the foreign are not more numerous than the native; and, though he is himſelf of opinion that they are not, it is evident that they muſt bear a conſiderable proportion to the whole *. In the petrifactions of Monte Bolca, near Verona, where the impreſſions of fiſh are preſerved between the laminæ of a calcareous ſchiſtus, one hundred and five different ſpecies have been enumerated, of which thirty-nine are from the Aſiatic ſeas, three from the African, eighteen from thoſe of South, and eleven from thoſe of North America †. Similar obſervations have been made on the marine plants, and the impreſſions of vegetables, found in rocks, in different parts of

 Europe.

* Buffon, Theorie de la Terre, art. 8.

† Sauſſure, Voyages aux Alpes, tom. iii. § 1535.

Europe. At St Chaumont, near Lyons, is found an argillaceous ſchiſtus, covering a bed of coal, every lamina of which is marked with the impreſſions of the ſtem, leaf, or other part of ſome plant; and it happens, ſays M. FONTENELLE, by an unaccountable deſtination of nature, that not one of theſe plants is a native of France. They are all ferns of different ſpecies, peculiar to the Eaſt Indies, or the warmer climates of America. Here alſo was found the fruit of a tree, which grows only on the coaſts of Malabar and Coromandel *.

The ſame holds of the bodies of amphibious animals which now make a part of the foſſil kingdom. The head and the bones of crocodiles have been found in the iſland of Shepey, at the mouth of the Thames; and the remains of an animal of the ſame ſpecies, but of a variety now peculiar to the Ganges, have been diſcovered in the alum rocks on the coaſt of Yorkſhire †. Theſe proofs of the tranſportation of materials

---

* Mém. de l'Acad. des Sciences, 1718, p. 3. and 287; and 1721, p. 89, &c.

† Phil. Tranſ. vol. l. p. 688. CAMPER denies that the remains here mentioned belong to the crocodile, or any amphibious animal, and refers them to the balæna. He paſſes the ſame judgment on thoſe foſſil bones from St Peter's Mount, near Maeſtrich, which have been ſuppoſed to belong

materials by the ſea, have the advantage of involving nothing hypothetical, and of being equally addreſſed to the geologiſts of every perſuaſion.

On this ſubject I cannot help obſerving, that the accurate compariſon of the animal exuviæ of the mineral kingdom, with their living archetypes, is not merely a curious inquiry, but is one that may lead to important conſequences, concerning the nature and direction of the forces which have changed, and are continually changing, the ſurface of the earth.

160. Theſe remarks I have thought it proper to add to the proofs of the compoſition of the preſent from former ſtrata, in order to ſhew, that the great tranſportation of materials involved in that ſuppoſition, is not only conformable to the hypotheſis of the Neptuniſts concerning the ſecondary ſtrata, but is alſo proved by the moſt direct evidence, independently of all hypotheſis. All this reaſoning regards the ancient ſtate of

 the

belong to the crocodile; he looks on them as belonging to whales, though of an unknown ſpecies. In this Mount, ſo famous for its petrifactions, he finds many ſpecimens of bones, which he thinks belong to the turtle. Phil. Tranſ. vol. lxxvi. p. 443. The opinion of an author, ſo well ſkilled in comparative anatomy, muſt be regarded as of great weight: if it takes from our argument in one part, it adds to it in another, and the acquiſition of the turtle makes up abundantly for the loſs of the crocodile.

the globe. Whether ſuch a travelling of ſtony bodies makes a part of the ſyſtem now actually carrying on, will be conſidered in another place *.

Note vi. § 13.

Mr Kirwan's notion of precipitation.

161. The Neptuniſt who has provided the means of diſſolving the materials of the ſtrata, has only performed half his work, and muſt find it a taſk of equal difficulty to force this powerful menſtruum to part with its ſolution. Mr Kirwan, aware in ſome degree of this difficulty, has attempted to obviate it in a very ſingular way. Firſt, he aſcribes the ſolution of all ſubſtances in water, or, in what he calls the chaotic fluid, to their being finely pulveriſed, or created in a ſtate of the moſt minute diviſion. Next, as to the depoſition, the ſolvent being, as he acknowledges, very inſufficient in quantity, the precipitation took place, (he ſays), on that account the more rapidly.

If he means by this to ſay, that a precipitation without ſolution would take place the ſooner the more inadequate the menſtruum was to diſſolve the whole, the propoſition may be true; but

* See Note xix.

but will be of no uſe to explain the cryſtalliza-tion of minerals, (the very object he has in view), becauſe to cryſtallization, it is not a bare ſubſi-dence of particles ſuſpended in a fluid, but it is a paſſage from chemical ſolution to non-ſolu-tion, or inſolubility, that is required.

If, on the other hand, he means to ſay, that the ſolution actually took place more quickly, and was more immediately followed by precipitation, becauſe the quantity of the menſtruum was in-ſufficient, this is to aſſert, that the weaker the cauſe, the more inſtantaneous will be its effect.

Of two propoſitions, the one of which is nu-gatory, and the other abſurd, it is not material to inquire which the author had in view.

Note vii. § 16.

Compreſſion in the mineral regions.

162. It is worthy of remark, that the effects aſcribed to compreſſion in the Huttonian Theo-ry, very much reſemble thoſe which Sir Isaac Newton ſuppoſes to be produced in the ſun and the fixed ſtars by that ſame cauſe. "Are not," ſays he, "the ſun and fixed ſtars great earths, vehemently hot, whoſe heat is conſerved by the greatneſs of the bodies, and the mutual action

 and

and reaction between them, and the light which they emit; *and whose parts are kept from fuming away, not only by their fixity, but also by the vast weight and density of the atmospheres incumbent upon them, and very strongly compressing them* *."

163. The fact, of water boiling at a lower temperature under a less compression, is sufficient to justify the supposition, that bodies may be made by pressure to endure extreme heat, without the dissipation of their parts, that is, without evaporation or combustion. A further *postulatum* is introduced in Dr Hutton's theory, namely, that compound bodies, such as carbonat of lime, when the compression prevents their separation, may admit of fusion, notwithstanding that the fixed part may be infusible when separated from the volatile. This assumption is supported by the analogical fact of the fusion of the carbonat of barytes, as mentioned in the text.

164. In a region where the action of heat was accompanied with such compression as is here supposed, there could be no fire, properly so called, and no combustion: this is admitted by Dr Hutton, and it is therefore a fallacious argument which is brought against his theory, from the impossibility of fire being maintained in

---

* Newton's Optics, Query 11.

in the bowels of the earth. This impoſſibility is preciſely what he ſuppoſes; and yet Mr Kirwan's arguments are directed, not againſt the exiſtence of heat in the interior of the earth, but againſt the exiſtence of burning and inflammation.

After taking notice *, that Sauſſure had ſucceeded, though with extreme difficulty, in melting a particle of limeſtone, ſo ſmall as to be viſible only with a microſcope, " what (adds he) muſt have been the heat neceſſary to melt whole mountains of this matter? Judging by all that we at preſent know of heat, ſuch a high degree could only be produced by the pureſt air, acting on an enormous quantity of combuſtible matter. Now, EHRMAN obſerved, that the combuſtion of two hundred and eighty cubic inches of air, acting on charcoal, was not able to effect the fuſion of one grain of Carrara marble; from whence it is apparent, that all the air in the atmoſphere, nor in ten atmoſpheres, would not melt a ſingle mountain of this ſubſtance, of any extent, even if there were a ſufficient quantity of inflammable matter for it to act upon. Judging alſo of ſubterraneous heat by what we know of that of volcanoes, no ſuch heat exiſts: the higheſt they in general produce, is that requiſite for the fuſion of the volcanic glaſs called

 obſidian,

* Geol. Eſſays, p. 453.

obſidian, which Sauſſure found not to exceed 115° of Wedgewood; but baſaltine, which requires 140° of Wedgewood, is never melted in the lavas of Ætna. How little capable, then, would volcanic heat be to effect the fuſion of Carrara marble, which, according to the ſame excellent author, would require a heat of upwards of 6300° of Wedgewood, if this pyrometer could extend ſo far? And in what circumſtances does Dr Hutton ſuppoſe this aſtoniſhing heat to have exiſted, and even ſtill to exiſt, under the ocean, in the bowels of the earth, where neither a ſufficient quantity of pure air, nor of combuſtible matter, capable of ſuch mighty effects, can, with any appearance of probability, be ſuppoſed to exiſt; and, without theſe, ſuch degrees of heat cannot even be imagined, without flying into the region of chimeras."

165. Now, this reaſoning is not applicable to Dr Hutton's hypotheſis of ſubterraneous heat, becauſe it is grounded on experiments, where that very ſeparation of the volatile and fixed parts takes place, which is excluded in that hypotheſis. When limeſtone or marble is expoſed to ſuch heat as is here mentioned, or even to heat of a degree vaſtly inferior, the carbonic gas is expelled, and the body is reduced to pare lime; from the refractory nature of which, as we learn from the fact relative

to

to barytes, mentioned above, no conclufion can be drawn as to the infufibility of the fame fubftance, when combined with the carbonic gas. The Carrara marble may require a heat of 6300° of Wedgewood, to melt it in the open air, where the carbonic gas efcapes from it; but under fuch a preffure as would retain this gas, it cannot be inferred, that it might not melt with the heat of a glafs-houfe furnace. In like manner, it may be true, that two hundred and eighty cubic inches of air, acting on charcoal, cannot effect the fufion of one grain of this marble, after its fixed air is driven off from it; but we cannot from thence draw any inference, applicable to a cafe where the carbonic gas is retained, and where the action of heat is independent of atmofpheric air.

Nothing, therefore, can be more inconclufive than this reafoning, as it proceeds on the fuppofition, that Dr Hutton's fyftem admits propofitions, which in fact it exprefsly denies.

166. Of the production and maintenance of heat, in circumftances fo different from thofe of ordinary experience, we can hardly be expected to give any explanation; but we are not entitled, merely on that account, to doubt of the exiftence of fuch heat. Mr Kirwan thinks otherwife: "Judging," he fays, "from all we at prefent know of heat, fuch a high degree of it, (as will melt limeftone), could only be produced by the pureft

reſt air, acting on an enormous quantity of combuſtible matter. Without theſe, ſuch degrees of heat cannot even be imagined, without flying into the region of chimeras *."

Now, in the firſt place, the high degree of temperature which is here underſtood, is probably not neceſſary to the purpoſes of mineralization, as has juſt been ſhewn; and, in the ſecond place, it is not FIRE, in the uſual ſenſe of the word, but HEAT, which is required for that purpoſe; and there is nothing *chimerical* in ſuppoſing, that nature has the means of producing heat, even in a very great degree, without the aſſiſtance of fuel or of vital air. Friction is a ſource of heat, unlimited, for what we know, in its extent, and ſo perhaps are other operations, both chemical and mechanical; nor are either combuſtible ſubſtances, or vital air, concerned in the heat thus produced. So alſo the heat of the ſun's rays in the focus of a burning glaſs, the moſt intenſe that is known, is independent of the ſubſtances juſt mentioned; and, though that heat certainly could not calcine a metal, nor even burn a piece of wood, without oxygenous gas, it would doubtleſs produce as high a temperature in the abſence as in the preſence of that gas.

It

* Geol. Eſſays, p. 454.

It is true, that it is not by the ſolar rays that ſubterraneous heat is produced; but ſtill, from this inſtance, we ſee, that there is no incongruity in ſuppoſing the production of heat to be independent of combuſtible bodies, and of vital air. We are indeed, in all caſes, ſtrangers to the origin of heat: philoſophers diſpute, at this moment, concerning the ſource of that which is produced by burning; and much more are they at a loſs to determine, what upholds the light and heat of the great luminary, which animates all nature by its influence. If we would form any opinion on this ſubject, we ſhall do well to attend to the ſuggeſtions of that great philoſopher, who was hardly leſs diſtinguiſhed from others by his doubts and conjectures, than by his moſt rigorous and profound inveſtigations. "May not great, denſe, and fixed bodies, when heated beyond a certain degree, emit light ſo copiouſly, as, by the emiſſion and reaction of its light, and the reflections and refractions of its rays within its pores, to grow ſtill hotter, till it comes to a certain period of heat, ſuch as is that of the ſun? And, are not the ſun and fixed ſtars great earths, vehemently hot, whoſe heat is conſerved by the greatneſs of the bodies, and the mutual action and re-action between them and the light which they emit * ?"

167. Some

* Newton's Optics, *ubi ſuprà.*

167. Some recent experiments, ſeem to make the ſuggeſtions in this query applicable to an opaque body like the earth, as well as to luminous bodies, ſuch as the ſun and fixed ſtars. The radiation of heat, where there is no light, was firſt rendered probable by the experiments of M. PICTET of Geneva *; and the only objections to which the concluſions from thoſe experiments ſeemed liable, are removed by the late very important diſcoveries of Dr HERSCHEL †. From theſe it appears, that heat is capable of refraction and reflection, as well as light, ſo that it is not abſurd to ſuppoſe, that *the heat of great, denſe, and fixed bodies, may be conſerved by the greatneſs of the bodies, and the mutual action and reaction between them and the heat which they emit.*

The exiſtence of ſubterraneous heat is ſtill further rendered probable from the reſearches of MAIRAN, which tend to ſhew, that there is another ſource of terreſtrial heat beſides the influence of the ſolar rays ‡.

Whatever be the truth with regard to theſe conjectures, it is certain, that the firſt and original ſource of heat is independent of burning. Burning is an *effect* of the concentration of heat; and

* Eſſai ſur le Feu.

† Phil. Tranſ. 1800. p. 84.

‡ Mém. de l'Acad. des Sciences, 1765. p. 143.

and though, by a certain reaction, it has the power of continuing and augmenting that heat, it never can be regarded as its primary and material cause. When, therefore, we suppose a source of heat, independent of fire and of burning, we suppose what certainly exists in nature, though we are not informed of the manner of its existence, nor of its place, otherwise than from considering the phenomena of the mineral kingdom.

168. Lastly, we are not entitled, according to any rules of philosophical investigation, to reject a principle, to which we are fairly led by an induction from facts, merely because we cannot give a satisfactory explanation of it. It would be a very unsound view of physical science, which would induce one to deny the principle of gravitation, though he cannot explain it, or even though the admission of it reduces him to great metaphysical difficulties. If indeed a downright absurdity, or inconsistency with known and established facts, be involved in any principle, it ought not to be admitted, however it may seem calculated to explain other appearanes. If, for instance, Dr Hutton held, that combustion was carried on in a region where there was no vital air, we should have said, that he admitted an absurdity, and that a theory founded on such *postulata* was worse than chimerical. But, if the only thing imputable

imputable to him is, that, being led by induction to admit the fufion of mineral fubftances in the bowels of the earth, he has affumed the exiftence of fuch heat as was fufficient for this fufion, though he is unable to affign the caufe of it, I believe it will be found, that his fyftem only fhares in an imperfection, which is common to all phyfical theories, and which the utmoft improvement of fcience will never completely remove.

169. Thus, then, we are led, it muft be allowed, into the *region of hypothefis and conjecture*, but by no means into that of *chimeras*. Indeed, the reproach of flying into the latter region, may be faid to come but ill from one, who has trode fo often the *crude confiftence* of the chaos, and who delights to dwell beyond the boundaries of nature. By fojourning there long, it is not impoffible that the eye may become fo accuftomed to fantaftic forms, that the figures and proportions of nature fhall appear to it deformed and monftrous.

NOTE VIII. § 24.

Sparry ftructure of calcareous petrifactions.

170. When the fhells and corals in limeftone are quoted by mineralogifts, it is not always confidered

confidered in what ftate they are found. In general, they have a fparry ftructure, very different from that of the original fhell or coral, of which, however, they retain the figure with wonderful exactnefs, though probably fometimes altered in fize. Though fparry, they are often foliated, and preferve their animal, in conjunction with their mineral, texture. Now, this cryftallization is a mark of fome operation, quite different from any that can be afcribed to the water in which thefe bodies had their origin, and by which they were brought into their place. They were impervious to water; and it cannot be faid that their fparry ftructure has been derived from the percolation of that fluid, carrying new calcareous matter into their pores. We can account for the change produced in them, I think, only by fuppofing them to have been foftened by heat, fo as to permit their parts to arrange themfelves anew, and to affume the characteriftic organization of mineral fubftances.

All fhells have not the change effected on them that is here referred to; thofe in chalk, for inftance, retain very much their original form in all refpects. This is what we might expect from the very different degree of intenfity, with which the mineralizing caufe has acted on chalk, and on limeftone or marble. In general, it is in the hardeft and moft confolidated

ted limeſtone, that the marine objects are moſt completely changed into ſpar.

It would be exceedingly intereſting to examine, whether any of the phoſphoric acid remains united to ſhells of either of theſe kinds. We might moſt readily expect it to be united, in a certain degree, to the ſhells that are leaſt mineralized.

This experiment would enable us alſo to appreciate the force of Mr Kirwan's argument againſt the finer marbles, ſuch as the Carrara, containing ſhells *. This argument proceeds on an experiment, mentioned in the *Turin Memoirs* for 1789, from which it appears, that no phoſphoric acid is found in pure limeſtone; and its abſence, Mr Kirwan ſays, cannot be attributed to fuſion, as phoſphoric acid is indeſtructible by heat.

He calls this a demonſtration; but, in order to entitle it to that name, it will be neceſſary, firſt, to prove, that phoſphoric acid exiſts in thoſe limeſtones which evidently conſiſt of ſhells in a mineralized ſtate. If theſe are found without phoſphoric acid, it is evident that the preceding argument fails entirely. If they are found to contain that acid, it will then no doubt afford a probability, though not a demonſtration, that Carrara

* Geol. Eſſays, p. 458.

Carrara marble does not directly originate from ſhells.

That nature has ſome proceſs, by which the above acid is ſeparated from the earth of bones, and probably alſo from the earth of ſhells, is evident from the ſtate in which the bones are found in the caves of Bayreuth. Thoſe that are the moſt recent, and leaſt petrified, contain moſt of the phoſphoric acid. Where the petrifaction has proceeded far, that acid is not found.

171. Among many of the ſtrata, ſuch a fluidity has prevailed, as to enable ſome of the ſubſtances included in them to cryſtallize. Calcareous ſpar and ſiliceous cryſtals are often found in ſtratified rocks, forming veins of ſecretion, or lining cloſe cavities, included on all ſides by the uncryſtallized rock. In the inſtances of gneiſs, and many ſpecies of marble, almoſt the whole matter of the ſtratum is cryſtallized. This union of a ſtratified and cryſtallized ſtructure in the ſame ſubſtance, has a great affinity to that union of the cryſtallized with the *organic* ſtructure of ſhells and corals which has juſt been mentioned; and both are doubtleſs to be referred to the ſame cauſe.

Note ix. § 31.

Petroleum, &c.

172. According to the theory of coal laid down above, its two chief materials, charcoal and bitumen, being furniſhed by the vegetable and animal kingdoms, both of the land and of the ſea, have formed with one another a new combination, by the action of ſubterraneous heat; but have alſo, in ſome caſes, been ſeparated by that ſame action, where the degree of compreſſion neceſſary for their union, happened to be wanting. The carbonic part, when thus ſeparated from the bituminous, forms an infuſible coal, which burns without flame: the bituminous part, when ſeparated from the carbonic, is found in the various ſtates of naphtha, petroleum, aſphaltes, and jet.

The great reſemblance of infuſible or blind coal, to the reſiduum obtained by the diſtillation of bituminous coal; and again, the coincidence of the bitumens juſt named, with the volatile part, or the matter brought over by ſuch diſtillation, are ſtrong arguments in favour of this theory. The other facts in the natural hiſtory of coal, ſerve to confirm the ſame concluſion; but it muſt be confeſſed, that what we we know of the pure bitumens, except the circumſtance juſt mentioned, is of a more ambiguous nature, and may be reconciled with different theories. The drops of petroleum contained within the cavities of the limeſtone, mentioned at § 31., are however ſtrong facts in confirmation of Dr Hutton's opinions, and they are furniſhed by the ſubſtances purely bituminous. A careful examination would probably make us acquainted with others of the ſame kind, for limeſtone is very often the matrix in which petroleum and aſphaltes are contained. The greateſt mine of aſphaltes in Europe, that in the *Val de Travers*, in the territory of Neufchâtel, is in limeſtone, from which, though it in ſome places exudes, it is in general extracted by the application of heat. The ſtrata for ſeveral leagues are impregnated with bitumen; and, if examined with attention, would probably afford ſpecimens ſimilar to thoſe which have juſt been mentioned.

173. It is a general remark, that, where petroleum is found, on digging deeper, they come to aſphaltes; and, at a depth ſtill greater, they diſcover coal. This probably does not hold invariably; but it is certain, that moſt of the fountains of petroleum are in the neighbourhood of coal ſtrata. Petroleum and aſphaltes are found in great abundance in Alſace, in a bed of ſand, between two beds of clay or argillaceous

ſchiſtus, and the ſame country alſo affords coal *. This is true likewiſe of the foſſil-pitch of Coal-Brookdale; and of the petroleum found in St Katharine's well, near Edinburgh. Auvergne † contains abundance of foſſil-pitch, which exudes, in the warm ſeaſon, from a rock impregnated with it through its whole maſs. There are alſo coal ſtrata in the ſame country, not far diſtant.

A very ſatisfactory obſervation relating to this ſubject, has lately been communicated from a country, with whoſe natural hiſtory we were till of late entirely unacquainted. In the Burmha empire, petroleum is dug up in an argillaceous earth, from the depth of ſeventy cubits. This argillaceous earth, or ſchiſtus, lies under a bed of freeſtone; and under all, about one hundred and thirty cubits from the ſurface, is a bed of coal ‡.

174. In the petroleum lake of the Iſland of Trinidad, deſcribed *Phil. Tranſ.* 1789, the petroleum evidently exudes from the rock, and is collected in a variety of ſprings in the bottom, after which it hardens, and acquires the conſiſtency

---

* Encyclopédie, mot, *Aſphalte*.

† Voyage en Auvergne, par Legrand, tom. i. p. 351.

‡ Aſiatic Reſearches, vol. vi. art. 6. p. 130.

ency of pitch. The manner, therefore, in which petroleum exifts in the ftrata, is very confiftent with the idea of its having been introduced in the form of a hot vapour.

Even amber appears to have fome relation to coal. It is found in the unconfolidated earth in Pruffia and Pomerania; but I am not fure whether this earth is *travelled* or not. In the fame earth where the amber is found, there is often a mixture of coaly matter, which burns in the fire; it is apparently fibrous, and has been confidered as a kind of foffil-wood *.

Thefe circumftances make out a connection between the purer bitumens and ordinary coal; but do not, it muft be acknowledged, eftablifh any thing with refpect to the more immediate relation, fuppofed in this theory to exift between them and blind-coal. It is probable, indeed, that, to difcover any facts of that kind, the natural hiftory of both fubftances muft be more carefully examined; the natural hiftory of blind-coal, in particular, has hitherto been but little attended to.

175. A fact is mentioned by Mr Kirwan, which muft not be regarded as lefs valuable for being adverfe to this theory. It is, that neither petroleum, nor any foffil bitumen, is found in the vicinity of the Kilkenny coal, as might be ex-

 pected,

* Buffon, Hift. Nat. des Mineraux, tom. ii. p. 5.

pected, if that coal was deprived of its bituminous part by subterraneous distillation *. This, however, admits of explanation. Though a general connection, on the above hypothesis, might be expected between bitumens and infusible coal, we cannot look for it in every instance. The heat which drove off the bitumen from one part of a stratum of coal, may only have forced it to a colder part of the same stratum; and thus, in separating it from one portion of carbonic matter, may have united it to another. Blind-coal may therefore be found where no bitumen has been actually extricated. In like manner, bitumen may have been separated, where the coal was not reduced to the state of coak, as a part of the bitumen only may have been driven off, and enough left to prevent the coal from becoming absolutely infusible.

It should be considered too, if the bitumen was really separated, and forced, in the state of vapour, into some argillaceous or limestone stratum, that this stratum may have been wasted and worn away long ago, so that the bitumen it contained may have entirely disappeared. It does not therefore necessarily follow, that, wherever we find blind-coal, there also we should discover some of the purer bitumens.

NOTE

* Geol. Essays, p. 473.

NOTE X. § 37.

The height above the level of the ſea at which the marks of aqueous depoſition are now found.

176. We have two methods of determining the *minimum* of the change which has happened to the relative level of the ſea and land; or for fixing a limit, which the true quantity of that change muſt neceſſarily exceed. The one is, by obſerving to what height the regular ſtratification of mountains reaches above the preſent level of the ſea; the other is, by determining the greateſt height above that level, at which the remains of marine animals are now found. Of theſe two criterions, the firſt ſeems preferable, as the fact on which it proceeds is moſt general, and leaſt ſubject to be affected by accidental cauſes, or ſuch as have operated ſince the formation of the rocks. The reſults of both, however, if we are careful to ſelect the extreme caſes, agree more nearly than could have been expected.

177. The mountain Roſa, in the Alps, is entirely of ſtratified rocks, very regularly diſpoſed,

 and

and nearly horizontal *. The higheſt ſummit of this mountain is, by Sauſſure's meaſurement, 2430 toiſes, or 14739 Engliſh feet, above the level of the ſea, or lower than the top of Mont Blanc only by 20 toiſes, or 128 feet †. This is, I believe, the higheſt point on the earth's ſurface, at which the marks of regular ſtratification are certainly known to exiſt; for though, by the account of the ſame excellent mineralogiſt, Mont Blanc itſelf is ſtratified, yet, as the rock is granite, the ſtratification vertical, and ſomewhat ambiguous, it is much leſs proper than Mont Roſa for aſcertaining the limit in queſtion.

178. Again, in the new continent, we have an inſtance of ſhells contained in a rock, not much lower than the ſummit of Mont Roſa. This is one deſcribed by Don ULLOA, near the quickſilver mine of Guanca-Velica, in Peru. The height at which a ſpecimen of theſe ſhells, given by Ulloa to M. le GENTIL, was found, was $2222\frac{1}{3}$ toiſes, or 14190 feet Engliſh, above the level of the ſea ‡. This height agrees with the preceding, within 549 feet, a quantity comparatively ſmall.

179. The

* Voyages aux Alpes, tom. iv. § 2138.

† *Ibid.* § 2135.

‡ See Hiſt. Acad. des Sciences, 1770. Phyſ. Générale, No. 7.

179. The laſt of the facts juſt mentioned is curiouſly commented on by Mr Kirwan. As he has proved, he ſays, that the mountains higher than 8500 feet were all formed before the creation of fiſh, it follows, that the ſhells found at Guanca-Velica, muſt have been carried there by the deluge *. Now, without objecting to the proof here referred to, (though it ſeems very open to objection), it is ſufficient to remark, that, if the ſhells at Guanca-Velica were carried there by the deluge, or any other cauſe that operated after the formation of the rock of which the mountain conſiſts, they can make no part of that rock, but muſt lie, like other adventitious foſſils, looſe and detached on the ſurface, or at moſt externally agglutinated to the ſtone. This, however, is certainly not the fact ; for, in the account juſt quoted, we read, that Don Ulloa told M. le Gentil, " qu'il avoit détaché ces coquilles d'un banc fort épais." This ſeems plainly to indicate, that the ſhells were included in a bed of rock. But, granting that the expreſſion is a little ambiguous, on turning to the *Mémoires Philoſophiques* of the ſame author, the difficulty is completely removed, and it is made evident, that theſe ſhells are in fact integrant parts of the rock. " On voit dans ces montagnes-là, (about Guanca-

* Geol. Eſſays, p. 54.

Guanca-Velica, and particularly at that in which is the quickfilver mine), des coquilles entières, petrifiées et enfermées au milieu de la roche, que les eaux de pluie mettent à decouvert. Ces coquilles font corps avec la pierre; mais malgré cela, on remarque que la partie qui fut coquille, fe diftingue par la couleur, la ftructure, la qualité de la matière de tout autre corps pierreux qui l'enferme, et du maffif qui s'eft fixé entre les deux ecailles *," &c. He goes on to fay, that one can diftinguifh marks of thefe fhells having been worn, before they were included in the ftone.

180. Thus it appears, that whatever proof any foffil-fhell affords, that the rock in which it is found was formed under the fea, the fame is afforded by the foffil-fhells of Guanca-Velica; and we are, therefore, perfectly entitled to conclude, that the relative level of the fea and land has changed, fince the formation of the latter, by more than 14000 feet. The height affumed in § 37. is therefore much under the truth; and the water, for which the Neptunifts muft provide room in fubterraneous caverns, might very well have been ftated at more

* Mém. Philofophiques de Don Ulloa, Difcours xvi. vol. i. p. 364.

more than a five-hundredth part of the whole maſs of the earth.

Thus alſo the argument by which the Neptuniſts would connect the creation of fiſh with the beginning of the ſecondary mountains, falls entirely to the ground. Indeed, it is ſtrange that Mr Kirwan ſhould have ſuppoſed it poſſible, that the ſhells in queſtion were looſe and unconnected with the rock, and had continued ſo, ever ſince the deluge, in ſuch elevated ground, where the torrents wear and cut down the mountains with unexampled violence, and have hollowed out *Quebradas* ſo much deeper and more abrupt than the glens or vallies among other mountains. He had not, I believe, ſeen the paſſage I have quoted from Ulloa ; but the circumſtances did not warrant the ſhells in queſtion to be regarded as extraneous and adventitious foſſils. A geologiſt ſhould have known better than to ſuppoſe this poſſible. When we ſee VOLTAIRE aſcribing to accidental cauſes the tranſportation of thoſe ſhells which he had been told were often found among the Alps, we can excuſe in a Poet and a Wit, that ignorance of the facts in mineralogy, which concealed from him the extreme abſurdity of his aſſertion ; but when a Chemiſt or Mineralogiſt talks and reaſons in the ſame manner, we cannot conſider him as entitled to the ſame indulgence.

NOTE

Note xi. § 42.

Fracture and diſlocation of the ſtrata.

181. The greateſt part of the facts relative to the fracture and diſlocation of the ſtrata, belongs to the hiſtory of veins. The inſtances of *ſlips*, where no new mineral ſubſtance is introduced between the ſeparated rocks, are what properly belong to this place. The frequency of theſe, and their great extent, are well known where-ever mines have been wrought. In ſome of them no opening is left, but the ſlipped ſtrata remain contiguous; in other caſes, there is introduced an unconſolidated earth, often a clay, which may be ſuppoſed to have come from above, and very probably to have been carried down by the water. In ſome ſuch caſes, however, there are not wanting appearances, which ſhow the matter in the ſlip to have been forced up from below, as we find it to contain ſubſtances which could not have come from the ſurface *.

182. A

* Unconſolidated earth contained between the ſides of a rock that has ſlipped, is frequent in Cornwall, and is called a *Fleukan*.

182. A very remarkable fact of this kind occurred not long ago, in digging the Huddersfield canal in Yorkſhire; and a very diſtinct account of it is given in the *Philoſophical Tranſactions*, by the engineer who directed the work. In carrying a tunnel into the heart of a hill, the miners came to what is called in the deſcription a *fault*, *throw*, or *break*, or what we have here called a ſhift, which was filled with *ſhale* ſet on edge, mixed with ſofter earth, and in ſome places with ſmall lumps of coal. The fault or ſpace filled with theſe materials, was in general about four yards broad, and lay nearly in the direction of the tunnel, ſo that a conſiderable extent of it was viſible. Beſide the ſhale, it contained a *rib* of limeſtone, about four feet thick, which run parallel to the ſides of the *fault*, and about four feet from the ſouthern margin of it. On each ſide of this rib were found balls of limeſtone, promiſcuouſly ſcattered, and of various ſizes, from an ounce to one hundred pounds weight. The balls, when broken, were found to contain ſome pyrites near their edges; they were not perfectly globular, but flattened on the oppoſite ſides, and ſimilar to one another *. At the time when the account was written, about ſeventy yards of the *rib* had been diſcovered.

183. Now,

* Phil. Tranſ. 1796. p. 350.

183. Now, it is certain, that neither this rib of limeftone, nor the balls that accompanied it, can have come from above, as there is no limeftone within twenty miles of the place where they were found. They muft, therefore, have been forced up from below, and no doubt belong to fome limeftone ftrata, which lie there at a great depth under the furface. The length of this fragment of rock; which, from the account, one muft fuppofe to have been entire, conveys no mean idea, either of the intenfity or regularity of the force by which it was brought into its prefent fituation. In veins, it is not uncommon to meet with ftones that appear to have come from a greater depth: but this is probably the moft remarkable inftance of the fame phenomenon, which has appeared in a mere flip, and none, I think, can fpeak a language lefs liable to be mifunderftood.

184. I fhall here mention another mark of violent fracture, that has been obferved in rocks of breccia or pudding-ftone, which, though not of the fame kind with the preceding, and of a nature quite peculiar, belongs rather to this place than any other. In rocks of the kind, juft mentioned, it fometimes happens, that confiderable portions are feparated from one another, as if by a mathematical plane, which had cut right acrofs all the quartzy pebbles in its way.

way. None of the pebbles is drawn out of its ſocket, that is, out of the cement that ſurrounds it, but is divided in two with a very ſmooth and even fracture. The pebbles, in the inſtances which I have ſeen, were of quartz, and other ſpecies of primary and much indurated rock.

Lord WEBB SEYMOUR and I obſerved pudding-ſtone rocks, exhibiting inſtances of this ſingular kind of fracture, near Oban, in Argyleſhire, about three years ago. The phenomenon was then entirely new to us both; but I have ſince met with an inſtance of the ſame kind in Sauſſure's laſt work. As the fact is of ſo particular a kind, I ſhall ſtate it in his own words: The place was on the ſea-ſhore, near the little town of Alaſſio, between Nice and Genoa.

" En paſſant entre ces blocs de breche, j'admirai quelques-uns d'entr'eux, d'une grandeur conſidérable, et taillés en cubes, avec la plus parfaite régularité. Il y avoit ceci de remarquable, c'eſt que l'action de la peſanteur, qui avoit taillé ces cubes en rompant leurs couches, avoit coupé tous les cailloux des breches à fleur de la ſurface de la pierre, auſſi nettement que ſi c'eût été une maſſe molle qu'on eût tranchée verticalement avec un raſoir. Cependant parmi ces cailloux, la plupart calcaires, il s'en trouvoit de très durs, de petroſilex, par exemple, même de jade,

jade, qui étoient tranchées tout aussi nettement que les autres *."

185. This description is no doubt accurate, though it involves in it something of theory, viz. that the fracture was made by the weight of the stone. This may indeed be true; the operation probably belongs altogether to the surface, and is one with which the powers of the mineral regions are not directly concerned. The phenomenon, however, appears to me, on every supposition, very difficult to explain. In the specimen which I brought from Oban, the smallest pieces of stone are cut in two, as well as the largest. The consolidation and hardness of the mass are very great, and the connection of the different fragments so perfect, that it is no wonder the whole should break as one stone. But still, that the fracture should be so exactly in one plane, and without any shattering, is not a little enigmatical; if it is indeed a fracture, it must be the consequence of an immense impulse, very suddenly communicated.

Note

* Voyages aux Alpes, tom. iii. § 1371.

NOTE XII. § 43.

Elevation and inflexion of the ſtrata.

186. The evidence of the different formation of the primary and ſecondary ſtrata, and of the changes which the former have undergone, is beſt ſeen at the points where thoſe ſtrata come into contact with one another. Dr Hutton was not the firſt who obſerved theſe junctions, though the firſt who rightly interpreted the appearances which they exhibit. He has mentioned obſervations of this ſort by De Luc on the confines of the Hartz; by the author of the *Tableau de la Suiſſe*, at the paſs of Yetz; by Voight, in Thuringia; and Schreiber, at the mountain of Gardette *.

The leading facts to be remarked, are,

I. The vertical or very upright poſition of the primary or lower ſtrata.

II. The ſuperſtratification of the ſecondary, in a poſition nearly horizontal, ſo as to be at right angles to thoſe on which they reſt.

III. The interpoſition of a breccia between them; or, as happens in many caſes, the tranſition of the loweſt of the ſecondary beds into a

O breccia,

* Theory of the Earth, vol. i. from p. 410. to 453.

breccia, containing fragments ſometimes worn, ſometimes angular, of the primary rock.

This laſt is a phenomenon extremely general, and all our ſubſequent information confirms Dr Hutton's anticipations concerning it. "It will be very remarkable," he ſays, "if ſimilar appearances, (ſuch as thoſe of the breccia deſcribed by Voight), are always found upon the junction of the Alpine with the level countries *." Sauſſure, in a part of his work, not publiſhed when Dr Hutton wrote this paſſage, has atteſted the generality of the fact with reſpect to the whole Alps, from the Tyrol to the Mediterranean: "Un fait que l'on obſerve ſans aucune exception, ce ſont les amas de débris, ſous la forme de blocs, de breches, de poudingues, de grés, de ſable, ou amoncelés, et formant des montagnes, ou des collines, diſperſée ſur le bord exterieur, ou même dans les plains qui bordent la chaine des Alpes †."

This paſſage is perfectly deciſive as to the generality of the fact, that the Alps, from the Tyrol to the Mediterranean, are bordered all round by pudding-ſtones or breccias. At the ſame time, it is neceſſary to remark, that M. Sauſſure, by enumerating looſe blocks and ſand, along with pudding-ſtones, breccias and grit, confounds together things which are extremely different, and which

* Theory of the Earth, vol. i. p. 448.

† Voyages aux Alpes, tom. iv. § 2330.

which have had their origin at periods extremely remote from one another. The consolidated rocks of breccia, pudding-stone and grit, though they are indications of waste, have received their present character at the bottom of the sea: the loose blocks of stone, the sand and gravel, on the other hand, are the effects of the waste now going forward on the surface of the land, and are the materials out of which rocks of the three kinds just mentioned may hereafter be composed. If so skilful a mineralogist as Saussure is guilty of such inaccuracy, it must be ascribed to the confusion necessarily arising from the system which he followed, and not to his own want of discrimination.

187. The same phenomenon, of a breccia circumscribing the primary mountains, is met with in Scotland; and the Grampians, wherever they are bounded by secondary strata, whether on the south or north, afford examples of it. The breccia generally consists of the fragments of the primary rock, most commonly rounded, but sometimes also angular, united by a cement of secondary formation, and the whole disposed in horizontal beds. It was on the constancy of this accompaniment of the primary strata, and on the great quantity of highly polished gravel often included in these breccias, that Dr Hutton grounded the hypothesis of the double raising

up and letting down of the ancient ſtrata. See § 43.

188. As the ſpots where the primary and ſecondary rocks may be ſeen in contact with one another are of great importance in geology, and preſent to the ſenſes the moſt ſtriking monuments of the high antiquity and great revolutions of the globe, it may be uſeful to point out ſuch of them as have been obſerved in this iſland. To thoſe which Dr Hutton has deſcribed, I have a few more to add, the reſult of ſome geological excurſions, which I made in company with the Right Honourable Lord WEBB SEYMOUR, to whoſe aſſiſtance I have been much indebted in the proſecution of theſe inquiries.

189. The moſt ſouthern junction which we obſerved is at Torbay, where the ancient ſchiſtus which prevails along the coaſt, from the Land's End to that point, receives a covering of red horizontal ſtandſtone, the ſame which compoſes the greater part of Devonſhire. The ſpot where the immediate contact is viſible, is on the ſhore, a little to the ſouth of Paynton; and one circumſtance, which among many others ſerves to diſtinguiſh the different formation of the two kinds of rock, is, that the ſchiſtus, which is elevated here at an angle of about 45°, is full of quartz veins, which veins are entirely confined

fined to it, and do not, in as far as we could obſerve, penetrate into the ſandſtone, in a ſingle inſtance. It is probable, that on the north ſhore of the bay, the ſame line of junction is viſible: we ſaw it at Babicomb Bay, ſtill more to the northward.

190. From this place, the ſecondary ſtrata of different kinds prevail without interruption, along the coaſt of the Britiſh Channel, and of the German Ocean, as far as Berwick upon Tweed, and for ſome miles beyond it. The ſea-coaſt then interſects a primary ridge, the Lammermuir Hills, which traverſes Scotland from eaſt to weſt, uniting, near the centre of the country, with the metalliferous range of Leadhills, and afterwards with the mountains of Galloway. The ſection which the ſea-coaſt makes of the eaſtern extremity of this ridge, is highly inſtructive, from the great diſturbance of the primary ſtrata, and the variety of their inflexions. The junction of theſe ſtrata with the ſecondary, on the ſouth ſide, is near the little ſea-port of Eyemouth, but the immediate contact is not viſible.

On the north ſide of the ridge, the junction is at a point called the *Siccar*, not far from Dunglaſs, the ſeat of Sir James Hall, Baronet. By being well laid open, and diſſected by the working of the ſea, the rock here diſplays the relation between the two orders of ſtrata to great advantage. Dr Hutton himſelf has deſcribed this junction; *Theory of the Earth*, vol. i. p. 464.

191. From the point juſt mentioned, the ſecondary ſtrata continue as far as Stonehaven, where the ſouthern chain of the Grampian mountains is interſected by the ſea-coaſt. Here a great maſs of pudding-ſtone appears to lie on the primary ſtrata, but their immediate contact has not been obſerved.

192. Going along the coaſt toward the north, the next junctions which we ſaw were on the ſhore, one near Gardenſton, and another near Cullen, in Banff-ſhire. The latter is very diſtinct; it is about a mile to the weſtward of the rocks called *The Three Kings*, where a red ſandſtone, the lower beds of which involve much quartzy gravel, lies horizontally upon very regular, upright, and highly indurated ſtrata. Some of theſe ſtrata are micaceous, and others of the granulated quartz, mentioned in § 152.

193. This laſt is, I believe, the moſt northern junction which has been obſerved in our iſland. The weſtern coaſt furniſhes ſeveral more, which however are not all viſible. The line of ſeparation, between the primary ſchiſtus of the Grampians and the ſandſtone which covers it, is interſected at its weſtern extremity by the Frith of Clyde, not far from Ardencaple in Dunbartonſhire. The two kinds of ſtone can be

traced

traced within a few yards of each other, but not to the actual contact: the beds of sandstone nearest the schistus form as usual a breccia, loaded with fragments of the primary rock. The secondary rock, which begins here, continues for about fifty miles south, to Girvan in Ayrshire, where the primary schistus again rises up, but is not seen in contact with the secondary. It extends to the Mull of Galloway and the shores of the Solway Frith.

The Isle of Arran, however, not far distant from this part of the coast, contains a junction at its northern extremity, where secondary strata of limestone lie immediately on a primary micaceous schistus. This is described by Dr Hutton, and was the first phenomenon of the kind which he had an opportunity of examining *. The junction is visible but at one spot, and is not seen so distinctly as in some of the instances just mentioned; but the great quantity of pudding-stone near it, renders it more interesting than it would be otherwise. As the greater part of this little island is surrounded by secondary strata, other junctions might be expected to be visible.

194. On the coast of England and Wales, from the Solway Frith to the Land's End, though there are several alternations from secondary to primary

* Theory of the Earth, vol. i. p. 429.

mary ſtrata, I know not that any of them have been obſerved. At St Bride's Bay, in Pembrokeſhire, the primary and ſecondary ſtrata are ſeen very near their junction ; but the preciſe line I believe is not viſible. The coal pits in the ſecondary ſtrata, approach here within a few hundred yards of the primary. The ſecondary ſtrata which commence at this place, occupy both ſides of the Briſtol Channel, and meet the Corniſh ſchiſtus, which extends acroſs the north of Devonſhire to the Quantock Hills, in a line that may be looked for on the ſea-coaſt, ſomewhere between Watchett and Minehead.

195. Beſides the ſea-coaſt, the beds of rivers may be expected to afford information on this ſubject. To the inſtances I have mentioned, I have accordingly two others from the inland country to be added. One of them is from the river Jed, a little way above Jedburgh, where the ſecondary ſtrata are ſeen lying horizontally on the primary, a ſection of both being made by the bed of the river. The phenomena here are very diſtinct, and ſtrongly marked : Dr Hutton has deſcribed and repreſented them in a plate *. He has mentioned another junction, not far from this, which he ſaw in the Tiviot. Both theſe belong to the ſame primary ridge with the Siccar Point.

196. I

* Theory of the Earth, vol. i. p. 430. ; alſo plate 3.

196. I ſhall mention only one other, which was diſcovered by Lord Webb Seymour and myſelf, at the foot of the high mountain of Ingleborough, in Yorkſhire. As we went along the Aſkrig road from Ingleton, about a mile and a half from the latter, an opening appeared in the ſide of the hill, on the right, about one hundred yards from the road, formed by a large ſtone, which lay horizontally, and was ſupported by two others, ſtanding upright. On going up to the ſpot, we found it was the mouth of a ſmall cave, the ſtone lying horizontally, being part of a limeſtone bed, and the two upright ſtones, vertical plates of a primary argillaceous ſchiſtus. The limeſtone bed, which formed the roof of the cave, was nearly horizontal, declining to the ſouth-eaſt ; the ſchiſtus nearly vertical, ſtretching from north-weſt by weſt, to ſouth-eaſt by eaſt. The ſchiſtus, though cloſe in contact with the limeſtone, ſeemed to contain nothing calcareous, and did not efferveſce with acids in the ſlighteſt degree.

As this cave is at the foot of Ingleborough, a cold wind, 24° below the temperature of the external air, which iſſued from the mouth of it, might very well be ſuppoſed to come from the inmoſt receſſes of that mountain. Ingleborough, which conſiſts entirely of ſtrata of limeſtone and grit, nearly horizontal, and alternating with one another, riſes to the height of 1800 or 2000 feet

above

above the ſpot where we now ſtood. This, I believe, is the greateſt thickneſs of ſecondary ſtrata that has ever been obſerved incumbent on the primary, and it is therefore a geological fact highly deſerving of attention. The country all round, to a very great extent, is compoſed of limeſtone, with a few beds of grit interpoſed, and forming, beſide Ingleborough, ſome other high mountains, ſuch as Wharnſide and Pennigant, all reſting, it is probable, on the ſame foundation.

At the ſpot juſt deſcribed, no breccia appeared to be interpoſed between the primitive and ſecondary rock ; but we found a breccia at another point of the ſame junction, not far diſtant. This was at a caſcade, in the river Greata, called Thornton Force, about two miles and a half from the place juſt mentioned. The Greata here precipitates itſelf from a horizontal rock of limeſtone ; and, after a fall of about eighteen or twenty feet, is received into a baſon which it has worked out in the primary ſchiſtus. This ſchiſtus is in beds almoſt perpendicular ; it exactly reſembles that which has juſt been deſcribed, and ſtretches nearly in the ſame direction. On the ſouth ſide of the river a breccia was ſeen, lying upon the ſchiſtus, or rather, it might be ſaid, that the loweſt beds of limeſtone contained in them many rounded fragments of ſtone, which, on compariſon, parifon, refembled exactly the fchiftus underneath. The primary rock itfelf is here feven or eight hundred feet above the level of the fea.

The fame fchiftus, fomewhat lower down the valley, and nearer to Ingleton, appears in large quantities, and is quarried for flate. Here, however, the immediate junction of the limeftone and fchiftus does not appear.

I have dwelt longer on the defcription of thefe appearances than on any others of the fame kind, becaufe, from the great mafs of fecondary ftrata which here covers the primary, the circumftances are fuch as we cannot expect to fee very often exemplified.

197. The Lakes of Cumberland are much vifited by travellers; and it may be worth remarking, on that account, that, as the fite of thefe lakes is a patch of primary country, bounded on all fides by fecondary, fo, in the rivers that run from the lakes, fuch junctions as we are now treating of may be expected to be found. Under Dun-Mallet, on the fide toward Ulles Water, we obferved a breccia, which was in horizontal layers, and feemed to lie on the primary fchiftus, fo that the whole hill is perhaps a piece of more indurated breccia, or fecondary rock, which has refifted the wearing and wafhing down of the rivers better than the reft.

198. After afcertaining the fact of the difturbance of the ftrata, and their removal from their original

original poſition, it is of conſequence to inquire into the direction of the force by which theſe changes have been produced. Now, if the diſturbed or elevated ſtrata, were every where in planes, without bending or ſinuoſity, it might perhaps be hard to determine, whether that force had acted in the direction of gravity, or in the oppoſite. Either ſuppoſition would account for the appearances; and, as gravity is a known force, providing we can find ſome place fit to receive the matter impelled downward by it, its action would furniſh the moſt probable ſolution of the difficulty.

It is on this principle that the Neptunian ſyſtem proceeds, imagining, that certain great caverns or vacuities having been opened in the interior of the globe, a great part of the waters which formerly covered its ſurface, retired into them, and much of the ſolid rock alſo ſunk down at the ſame time. In this way, one extremity of a ſtratum has been elevated, while the other has been depreſſed, and a certain inclination to the horizon has been given to the whole of it. Thus one cauſe ſerves two purpoſes; the vacuities in the interior of the earth account, both for the depreſſion of the ſea, and the elevation of the land; and the Neptuniſts, if the phenomena were all ſuch as have been now ſtated, might boaſt of a felicity of explanation, not very uſual in their ſyſtem.

But

But this appearance of ſucceſs vaniſhes, when the elevation and diſturbance of the ſtrata are more minutely examined, and are found to include waving and inflexion, in a great variety of forms. It then becomes evident, that the beds of rock, at the time when they were diſturbed from their horizontal poſition, had not their preſent hardneſs and rigidity, but were, in a certain degree at leaſt, ſoft and flexible. Without theſe qualities, they could not have received, as they have often done, the curvature of a circle, not many feet, nay, not many inches, in diameter; nor could they have been bent into ſuperficies, with their curvature in oppoſite directions, ſo that the ſame ſurface is in one part convex, and in another concave, on the ſame ſide, with a line of contrary flexure interpoſed. Theſe are appearances, not reconcilable with the mere falling in, and breaking down of indurated rocks.

199. The inflexions and wavings that we are here ſpeaking of, though not peculiar to the primary ſtrata, are found moſt frequently among them, and are perfectly familiar to every one who has travelled among mountains with any view to the ſtudy of geology. The following are a few inſtances of this phenomena, out of a great number which might be produced.

Sauſſure,

Sauſſure, in deſcribing the rout from Geneva to Chamouny, mentions many remarkable inſtances of the bending of the ſtrata, and particularly where the ſmall ſtream of Nant d'Arpenaz forms a caſcade, by falling over the face of a perpendicular limeſtone rock. The ſtrata of this rock are bent into circular arches, extremely regular, and with their concavity turned to the left. What deſerves particularly to be remarked, is, that a mountain behind the caſcade has its ſtrata bent in a direction oppoſite to the former, or with their concavity to the right. There is no doubt that the ſtrata of both rocks are the ſame, ſo that a vertical ſection of them would give a curve, in the figure of an S *. Theſe circumſtances are mentioned by Sauſſure, and from them we may infer this other property of theſe ſtrata, that their ſection by a horizontal plane, muſt exhibit a ſyſtem of ſtraight lines, probably all parallel to one another.

The ſame mineralogiſt deſcribes the calcareous ſtrata which compoſe the mountain Axenberg, on the ſide of the Lake of Lucerne, as having from top to bottom of the mountain the form of the letter S compreſſed (*écraſée*), with their curvature in ſome places very great. Theſe inflexions

---

* Voyages aux Alpes, vol. i. § 472.; alſo, Theory of the Earth, vol. ii. p. 30.

ions are repeated ſeveral times, and often in contrary directions; the layers are ſometimes broken, where their curvature is greateſt *.

On the ſide of the ſame lake, is another inſtance of bent ſtrata, in a mountain, of which the beds are horizontal in the lower part, but are bent at one end upwards, in the form of the letter C. The horizontal part is of great extent, and the rock is alſo calcareous †.

The *Montagne de la Tuile*, near Montmelian, receives its name from the beds of rock being incurvated in form of a tyle ‡. Among ſecondary mountains, the ſame kind of phenomena are obſerved, though leſs frequently, and with leſs variety of inflexion. The chain of Jura is ſecondary, and the beds which compoſe it are of limeſtone, or of grit: they are bent in ſuch a manner, that in a tranſverſe ſection of the mountain, each layer would have the figure of a parabola §.

200. The Pyrenees furniſh abundance of phenomena of the ſame kind, as we learn from the *Eſſai ſur la Mineralogie des Pyrénées*. The calcareous

---

* Voyages aux Alpes, tom. iv. § 1935.

† *Ibid.* § 1937.

‡ *Ibid.* vol. iii. § 1182, and plate i.

§ *Ibid.* tom. i. § 334.

calcareous ſtrata of the valley of Aſpe, repreſented plate v. of that work, deſerve particularly to be remarked.

201. Our own iſland abounds with examples of the bending and inflexion of the ſtrata, eſpecially the primary, and many of them very much reſembling thoſe in the Alpes and Pyrenees. On the top of the mountain of *Ben-Lawers*, in Perthſhire, there is a rock, the face of which exhibits a ſection of a great number of thin equidiſtant layers, bent backwards and forwards, like thoſe deſcribed by Sauſſure; and this unequivocal proof of the rock having once exiſted in the ſtate of a flexible and tenacious paſte, is rendered more ſtriking, by the great elevation of the ſpot, and the ruggedneſs and induration, both of the ſtone itſelf, and of every thing that ſurrounds it. Many other mountains in this tract conſiſt of a ſchiſtus, which is talcoſe rather than micaceous, and ſubject, in a remarkable degree, to the ſort of ſinuoſity and inflexion here treated of.

The appearances of the primary ſtrata on the coaſt of Berwickſhire, have been already mentioned, as affording much valuable inſtruction in geology. They alſo exemplify the waving and inflexion of the ſtrata on a large ſcale, and with great variety. A ſection of ſome of them

is

is given by Dr Hutton, in his *Theory of the Earth*, vol. i., from a drawing made by Sir James Hall. The nature of the curve ſuperficies into which the ſchiſtus is bent, is the better underſtood from this, that, beſides tranſverſe ſections from north to ſouth, the deep indentures which the ſea has made, and the projecting points of rock, exhibit many longitudinal ſections, in a direction from eaſt to weſt.

202. The dock-yards at Plymouth are in ſeveral places cut out of a ſolid rock of primary ſchiſtus, ſingularly incurvated. The inflexions are ſeen there to great advantage, being exhibited in three ſections, at right angles to one another, tranſverſe, longitudinal and horizontal.

203. From theſe inſtances, to which it were eaſy to add many more, two concluſions may be drawn. The firſt of theſe is very obvious, viz. that the ſtrata muſt have been pliant and ſoft when they acquired their preſent form. The bending of an indurated bed of ſtone into an arch of great curvature, and without fracture, as in the preceding examples, is a phyſical impoſſibility. Sauſſure has indeed obſerved a fracture to accompany the bending, in one or two caſes; but it is an uncommon phenomenon, and, where it happens, muſt no doubt be underſtood to indicate an imperfect flexibility. Now, if it be granted that the ſtrata were at any time

ſoft and flexible, ſince their complete formation, it will be found impoſſible to deny their having been ſoftened by the application of heat.

204. The ſecond concluſion, alluded to above, reſults from a property, which belongs very generally, if not univerſally, to the inflexions of the ſtrata. This conſiſts in their curvature being ſimple, or in one dimenſion only, like a cylindric ſuperficies, not double, or in two dimenſions, like the ſuperficies of a ſphere or ſpheroid. This may be otherwiſe expreſſed, by ſaying, that the ſections of the bent ſtrata, by a horizontal plane, are ſtraight lines, parallel to one another. On this account, every ſuch ſtratum ſeems as if it were bent over an axis, and the axes of all theſe different bendings, for a great extent of country, are nearly parallel.

The truth of this is evident, where the ſtrata are ſeen both tranſverſely and longitudinally. It holds remarkably of the primary ſchiſtus on the coaſt of Berwickſhire; where the beds of rock, if cut tranſverſely, by a vertical plane, exhibit the figures of very complicated curves, with various *maxima* and *minima*, and points of contrary flexure; but, if they are cut by a horizontal plane, the ſection will produce nothing but ſtraight lines, nearly parallel.

205. The

205. The conſtancy of the direction of the primary ſtrata, when eſtimated by their interſection with the horizontal plane, is often very remarkable. Their elevation and flexure are ſubject to great and ſudden changes, ſo as to paſs not only from greater to leſs, but from one ſide to the oppoſite, within a ſmall diſtance; but the horizontal line in which they *ſtretch*, uſually preſerves the ſame bearing to a great extent. The general direction of the primary ſtrata, in the ſouth part of Scotland, is from E. N. E. to W. S. W.; and the ſame is nearly true of thoſe which compoſe the ridge of the Grampians on the north, and the hills of Cumberland and Weſtmoreland toward the ſouth, though between the ſchiſtus of theſe three tracts, there is no communication at the ſurface, each being entirely ſeparated from the one next it, by the interpoſition of ſecondary ſtrata. I have already mentioned the obſervations of Lord Webb Seymour and myſelf, at the foot of Ingleborough; and it appears from them, that the vertical ſchiſtus on which that mountain reſts, though it ſtill preſerves an eaſtern and weſtern direction, varies ſeveral points from that of the more northern ſtrata. The ſtrata of Wales return more to the firſt-mentioned direction, and thoſe of Devonſhire and Cornwall agree with it very nearly. In all this, it will be eaſily con-

 ceived,

ceived, that I do not mean to ſpeak with abſolute preciſion, or to deny the exiſtence of great local irregularities. The reſult given is only a kind of average, deduced from obſervations hardly ſuſceptible of great exactneſs, and not yet ſufficiently multiplied to give to the concluſion all the accuracy it may attain.

206. This tendency of the primary ſtrata to take a uniform direction, has alſo been obſerved in other countries. Sauſſure remarked in the Alps, that the beds of ſchiſtus are generally parallel to the chains of mountains compoſed of them *; and this remark is probably applicable to all mountains conſiſting of primary ſtrata. The general direction, therefore, of the ſchiſtus of the Alps, muſt be confined between W. 10° S. and W. 40° S. In the Pyrenees, the direction of the ſtrata is about W. N. W †. If Sauſſure's rule may be depended on, the ſchiſtus of the Altaic, and moſt of the other great chains in the old continent, are in directions that run conſiderably to the ſouth of weſt. The Ourals, and perhaps ſome other of the northern chains, are however entirely different. In the Ourals, as we learn not only from the general direction of the chain, but from a ſection of it in the 10th volume

* Voyages aux Alpes, tom. i. § 577.

† Eſſai ſur le Mineralogie des Pyrenées.

lume of the Nova Acta of Petersburgh, (tab. 12.), the direction of the strata is nearly from N. to S. This last is probably the direction in the great chains of South America; so that the uniformity of direction in the primary strata, which some mineralogists would extend to those of the whole earth, is certainly imaginary, though there can be no doubt that it extends over very large portions of the earth's surface *.

207. The

* It is perhaps unnecessary to observe, that the two propositions, that the intersections of the strata with the horizon are parallel lines; and that they are lines which preserve the same bearing with respect to the points of the compass; are nearly the same thing for tracts of moderate extent, but for large portions of the earth's surface are extremely different. If, for instance, the belt of primary vertical schistus, which traverses the south of Scotland, were to be produced eastward in the same plane, from its northern extremity, where its direction is E. N. E. and its latitude 55°.57′, it would cut the meridian always less obliquely as it advanced, till, having increased its longitude about 26°.28′, it would be at right angles to the meridian, and its direction of consequence due east and west. This would happen in the parallel of 58°.51′, (on the shore of the Gulf of Finland, near Revel), the strata being now extended about 880 G. miles from the Siccar Point. Conversely, vertical strata, having the same bearing with respect

207. The tendency of the primary ſtrata to remain ſtraight in the horizontal direction, and to be bent in the vertical, is a phenomenon which points very directly to the cauſes from whence it has ariſen. A ſurface of ſimple curvature, or a ſurface ſtraight in one direction, is what

---

reſpect to the meridian, may be in planes very much inclined to one another. A ſtratum which bears eaſt and weſt in Cornwall, and one that does the ſame at the eaſt end of the Altaic chain, will be in planes, which, if produced, would cut one another at right angles. All this is ſufficiently plain from the doctrine of the ſphere, and is mentioned here, merely as a caution to prevent too haſty concluſions from being drawn from any correſpondence of bearing among the ſtrata of remote countries.

For the ſake of thoſe who would deduce the medium bearing of any body of ſtrata from a number of obſervations, it may be proper to take notice, that the true average is not to be found by ſimply taking an arithmetical mean among all the obſervations. A more exact way is to work by the traverſe table, as in keeping a ſhip's reckoning, (ſuppoſing the diſtance run to be always unity), and to compute from the obſerved bearings the amount of all the ſouthing or northing, and alſo of all the eaſting or weſting. The ſum of all the latter, divided by the ſum of all the former, is the tangent of the angle which the general direction of the ſtrata makes with the meridian.

what the application of forces to different points of a plane, which is flexible, though with a certain degree of rigidity, will naturally produce. The ſuppoſition, therefore, that theſe ſtrata were once flat and horizontal, and were impelled upward from that ſituation before they had become rigid or hard, will explain their having the kind of curvature which removes them as little as poſſible from their original condition. But no other hypotheſis affords any reaſon why they ſhould have that curvature more than any other. From the falling in of roofs of caverns, we might expect fracture and diſlocation, without any order or regularity ; but certainly no bending or ſinuoſity, nor any ſymmetrical arrangement. If, as ſome mineralogiſts allege, the curvature, as well as inclination of the ſtrata, aroſe from the irregularities of the bottom on which they were depoſited, why is the former in one dimenſion only, and why is it not in every direction, like that of hills and valleys, or the actual ſurface of the earth ? Or, laſtly, if the whole ſtructure of the primitive mountains is an effect of cryſtallization, and if theſe mountains are now ſuch as they have ever been from the time of their conſolidation, whence is it, that, in their bendings the law juſt mentioned is ſo conſtantly obſerved ? Indeed, the idea of aſcribing the inflexions of the ſtrata to cryſtallization, though ſuggeſted

gested by Saussure *, and since become a favourite system with several mineralogists, appears to me in the highest degree unsatisfactory and illusive. The purpose for which crystallization is here introduced, is not to give a specific figure to a particular substance, but to arrange the substances which it has formed and figured, according to certain rules; a work which we know not how it is to perform, and in which we have no experience of its power. Accordingly, this principle does not account, in any way whatever, for the circumstances which attend the inflexion of the strata, for the simple curvature which they affect, nor for that parallelism of their layers, which, in all their bendings, is so accurately preserved. It does, indeed, so little serve to explain these facts, that, were the appearances completely reversed; did the strata assume the most complex, instead of the most simple curvature; instead of equidistant, were they converging, or alternately receding and approaching to one another; the theory of crystallization might be equally applied to them. The state of the phenomena is a matter of perfect indifference to such a theory as this: all things are explained by it with the same facility; the straight and the

* Voyages aux Alpes, tom. i. § 475.

the crooked, the ſquare and the round, the moveable and the immoveable. Is it not evident that ſuch an explanation is a mere word ; or, if any thing more than a word, an expreſſion of our ignorance, ſo awkward and indirect, as to deprive us of whatever credit might have been gained by a plain and candid avowal of it ?

It ſhould never be forgotten, that a theory which accounts for *any thing*, and a theory which accounts for *nothing*, ſtand preciſely on the ſame footing, and ought to be baniſhed from all parts of philoſophy, as they have been from thoſe ſciences which are juſtly honoured with the name of accurate. The animated orbs of Ariſtotle, and the vortices of Des Cartes, have long ceaſed to be mentioned in phyſical aſtronomy ; the firſt, becauſe they accounted for every thing alike ; the ſecond, becauſe, when they accounted for one thing, they never could be made to account for another. Both theories, therefore, have very properly been rejected ; and, when geology ſhall undergo a ſimilar purification, the principle we have been conſidering will not be the only ſacrifice required of the Neptunian ſyſtem.

208. An appearance obſerved in ſome kinds of primary ſchiſtus, which clearly indicates their depoſition by water, and in planes very different from thoſe in which we now ſee them, though it might have been introduced before, is alſo

much

much connected with the prefent argument. This appearance confifts of fmall wavings or undulæ on the furface of the plates of fchiftus, precifely fimilar to thofe marks which are left by the fea on a gently inclining beach of fand, at the ebbing of the tide. All the fpecies of fchiftus do not feem to afford inftances of thefe wavings. The rocks which do fo, are, I think, chiefly of the argillaceous kind, but often highly indurated; fo that the laminæ containing the impreffions are not to be torn afunder but with great difficulty. Inftances of it abound in the fchiftus of Berwickfhire, and are alfo not unfrequent in that of Galloway. All muft agree about the agent which produced thefe marks; it could be no other than the fea; but it muft have been the fea acting on loofe, fmall and round particles, lying on a furface which was nearly horizontal.

209. Dr Hutton's theory is no where ftronger, than in what relates to the elevation and inflexion of the ftrata; points in which all others are fo egregioufly defective. The phenomena to be connected are here extremely various, and even in appearance contradictory: the horizontality of one part of the ftrata; the inclined or vertical pofition of another; the perfect planes in which one fet are extended; the breaking and diflocation

diſlocation found in a ſecond ; the inflexion and ſinuoſity of a third ; and almoſt every where the utmoſt rigidity and induration, combined with appearances of the greateſt ſoftneſs and flexibility ; the preſervation of a paralleliſm of ſuperficies in the midſt of ſo much irregularity, and the aſſumption of a determinate ſpecies of curvature, under circumſtances the moſt diſſimilar ; all theſe appearances were to be connected with one another, and with the conſolidation of the ſtrata, and this is done by the twofold hypotheſis, of aqueous depoſition, and the action of ſubterraneous heat. When theſe circumſtances are fairly conſidered, and when the ſhifts which other ſyſtems are put to on this occaſion are remembered, I think it will be granted, that few attempts at generalization have been more ſucceſsful, than that which is here made by the Huttonian Theory.

210. To the fact of the elevation of the ſtrata, the ſtudy of geology is much indebted. The ſtratified form of a great proportion of the earth's ſurface, gives to minerals that organization and regularity, which makes their diſpoſition an object of ſcience, and their inclined poſition ſerves to bring that organization into view, from far greater depths than we can ever reach by artificial excav tions. If, for inſtance, the termination of ſtrata, that make with the horizon

an

an angle of 30°, lying one over another, is ſeen for a horizontal diſtance of two miles; then it is certain, that if theſe ſtrata have that extent under ground, which may be reaſonably ſuppoſed, the thickneſs of the whole maſs, meaſured by a line perpendicular to its ſtratification, is half the horizontal diſtance, or amounts to one mile. It would alſo require a pit to be ſunk from the uppermoſt of theſe ſtrata, to the depth of (2 miles × tan 30°, =) 6093 feet, before it could interſect the undermoſt; and therefore, if we ſuppoſe the ſame ſtratum to preſerve the ſame character for the extent of ſome miles, we obtain the ſame information from inſpecting the edge-ſeams, and ſee in reality as far into the bowels of the earth, as if we had ſunk a perpendicular ſhaft to the depth of 6000 feet.

In general, the length of the horizontal line drawn acroſs the ſtrata, from the loweſt in poſition to the higheſt, multiplied into the ſine of the inclination of the ſtrata to the horizon, gives the thickneſs of the whole, meaſured perpendicularly to the plane of the ſtratification: and the ſame horizontal diſtance, multiplied into the tangent of the inclination, gives the actual depth at which the loweſt ſtratum would meet a perpendicular to the horizon, drawn from the higheſt extremity of the upper ſtratum.

In

In many cafes, the extent of ftratified materials admitting of fuch an examination as this, is much greater than has now been fuppofed. M. Pallas defcribes a range of hills on the fouth-eaft fide of the peninfula of the Tauride, which is cut down perpendicularly toward the fea, and offers a complete fection of the parallel beds of a primary, or, as he calls it, an ancient limeftone, inclined at an angle of 45° to the horizon; and this fection continues for the length of 130 *verfts*, or about 86 Englifh miles. The beds are fo regular, that M. Pallas compares them to the leaves of a book*. The height of thefe hills does not exceed 1200 feet, but the real height of the uppermoft ftratum above the undermoft, is $86 \times \sqrt{\frac{1}{2}} = 86 \times \frac{5}{7} = 61$ miles nearly.

If therefore we conceive that there is no fhift in all this great fyftem of ftrata, we in reality are enabled, by means of it, to fee no lefs than 61 miles into the interior of the earth, nearly a 65th part of the radius of the globe. It is true, that we can hardly fuppofe fo great a body of ftrata to have been raifed without fhifting, fo that we muft diminifh this depth confiderably; but were it reduced even to one-half, it will

* See Nova Acta Acad. Petropol. tom. x. (1792,) p. 257.

will appear, that men ſee much farther into the interior of the globe than they are aware of, and that geologiſts are reproached without reaſon for forming theories of the earth, when all that they can do is but to make a few ſcratches on its ſurface. Art indeed can do little more; but nature ſupplies the deficiency, and makes diſcoveries to the attentive obſerver, on the ſame great ſcale with her other operations.

The ſimpleſt account that can be given of the vaſt body of parallel and highly inclined ſtrata juſt mentioned, is, that it conſiſts of the ends of horizontal ſtrata, or of ſtrata not greatly inclined. that have been forced up when they were all ſoft and flexible. This is a much more conceivable ſuppoſition than Pallas's, viz. that the greater part of this maſs has ſunk down into ſome vaſt cavern in the interior of the earth.

NOTE

Note xiii. § 53.

Metallic Veins.

211. The large ſpecimens of native iron found in Siberia and Peru, mentioned above, § 51., are among the moſt curious facts in the natural hiſtory of metals. It has been doubted, however, by ſome, whether they really belong to natural hiſtory, or are not rather to be accounted artificial productions. If they had been found in the heart of rocks, or in the midſt of metallic veins, no doubt of this ſort could poſſibly have been entertained; but, as they lie quite on the ſurface, in the middle of flat countries, and at a diſtance from any known vein of metal, the conjecture that they may be artificial, and the remains of the iron founderies of ancient and unknown nations, is at firſt ſight not entirely deſtitute of probability. This probability, however, will appear to be the leſs, the more carefully the ſpecimens are examined. The metal is too perfect, and the maſſes too large, to have been melted in the furnaces, or to have been tranſported by the machinery, of a rude people. The ſpecimen in South America weighs 300 quintals, or about 15 tons, and is ſoft

ſoft and malleable *. The Siberian ſpecimen, deſcribed by Pallas, is alſo very large; it is ſoft and malleable, and full of round cavities, containing a ſubſtance, which, on examination, has been found to be chryſolite †. Now, it is certainly quite impoſſible, that, in an artificial fuſion, ſo much chryſolite could have come by any means to be involved in the iron; but, if the fuſion was natural, and happened in a mineral vein, the iron and the chryſolite were both in their native place, and their meeting together has nothing in it that is inexplicable.

212. Some circumſtances in the deſcription of the ſpecimen in South America, ſuch as the impreſſions of the feet of men and of birds on its ſurface, are not to be accounted for on any hypotheſis, and certainly require more careful inveſtigation. It is ſaid, that this iron is very little ſubject to ruſt, and the analyſis of a piece of it by PROUST makes it probable, that it owes this quality to its union with nickel ‡. It appears, alſo, that the country of Chaco, where this ſpecimen was found, affords many others of the ſame kind, one of which is mentioned in the deſcription above referred to. That country

* Phil. Tranſ. 1788. p. 37. alſo p. 183, &c.

† Kirwan's Mineralogy, vol. ii. art. Native Iron.

‡ Annales de Chimie, tom. xxxv. Meſſidor, p. 47.

country lies on the eaſt ſide of the Plata, and is a plain, extremely level, and of vaſt extent, without any appearance of mineral veins; but ſuch veins may nevertheleſs exiſt undiſcovered, in a tract ſubject to periodical inundations, and where the native rock is covered with alluvial earth and gravel to a great depth. The veins may be waſhed away, and the more durable ſubſtances, ſuch as thoſe pieces of native iron, may be left behind; and, though they muſt be of a formation extremely ancient, according to this hypotheſis, they may not have been very long on the ſurface.

213. Specimens of native iron have been found, leſs remarkable than the preceding for their ſize, but in circumſtances that excluded all idea of artificial fuſion. Of this ſort was MARGRAAF's ſpecimen of native iron, the firſt of the kind that was known; it conſiſted of ſmall bits of ſoft and malleable iron, found in the heart of a brown iron-ſtone*. This makes it certain, that native iron is a natural production, and the mere circumſtance of great magnitude, in the ſpecimens before mentioned, does not entitle us to doubt of their having that ſame origin. It is a circumſtance, beſides, not in the leaſt material to this argument; the ſmalleſt

 piece

* Kirwan's Mineralogy, vol. ii. p. 156.

piece of native iron being as much a proof of fusion as the greatest; and the specimen of Margraaf being just as conclusive in favour of the Huttonian Theory, as those of Pallas or De Celis, supposing their reality as mineral productions to be completely established. A metal malleable and ductile, in ever so small a quantity, cannot be the result of precipitation from a menstruum, without a very particular combination of circumstances. Such a metal, on the other hand, can be readily produced by igneous fusion; so that here the negative and affirmative parts of the inductive argument may both be regarded as complete.

214. Mr Kirwan, in order to account for the magnitude of the two large specimens mentioned above, supposes, that small pieces of native iron (about the formation of which he appears to have no difficulty), have been originally agglutinated by petroleum, and left bare, when the surrounding stony or earthy masses either withered or were washed off*. This is no doubt the most singular of all the opinions which have been advanced on the subject; and, as it borrows nothing from analogy, it admits of no proof, and requires no refutation. None but a chemist of eminence could have ventured with impunity

* Geol. Essays, p. 405.

impunity on an aſſertion ſo inconſiſtent with all the phenomena and principles of his ſcience.

215. A remark of the ſame author, on the ſubject of the native gold found in the county of Wicklow in Ireland, is entitled to more attention. "That theſe lumps of native gold," he ſays, "were never in fuſion, is evident from their low ſpecific gravity, and the grains of ſand found in the midſt of them. I found the ſpecific gravity of a lump of the ſize of a nutmeg to be only 12800, whereas, after fuſion, it became 18700 *."

This argument is plauſible; but, I think, nevertheleſs inconcluſive. The ſand found in the gold, accounts, at leaſt in part, for its lightneſs. It is only by repeated fuſions that any of the metals is brought to its utmoſt purity and higheſt ſpecific gravity; and on no ſuppoſition can the melting of gold in the mineral regions, be very likely to ſeparate it from heterogeneous ſubſtances. That quartzy ſand ſhould be found in it, after ſuch a proceſs, is naturally to be expected. The impreſſions which the quartz cryſtals have left on the Wicklow gold, would be received as a full proof of the fuſion of that metal, if geologiſts always regu-

 lated

* Geol. Eſſays, p. 402.

lated their theories by the principles which determine the belief of ordinary men.

216. Don Rubin de Celis, in the paper referred to above, mentions ſome maſſes of ſilver found at Quantajaia, and alſo ſome duſt of platina, in terms that excite a ſtrong deſire to have more information concerning them. They are conſidered by him as effects of volcanic fire; ſo we may conclude, that they contain evident marks of fuſion, and would in this ſyſtem be aſcribed to that heat, from which volcanic fire is but a partial and accidental derivation.

217. The ſtate alſo in which gold and ſilver are often found pervading maſſes of quartz, and ſhooting acroſs them in every direction, furniſhes a ſtrong argument for the igneous origin, both of the metal and the ſtone. From ſuch ſpecimens, it is evident, that the quartz and the metal cryſtallized, or paſſed from a fluid to a ſolid ſtate, at the ſame time; and it is hardly leſs clear, that this fluidity did not proceed from ſolution in any menſtruum: For the menſtruum, whether water or the *chaotic fluid*, to enable it to diſſolve the quartz, muſt have had an alkaline impregnation; and, to enable it to diſſolve the metal, it muſt have had, at the ſame time, an acid impregnation. But theſe two oppoſite qualities could not reſide in the ſame ſubject; the acid and alkali would unite together, and,

if

if equally powerful, form a neutral ſalt, (like ſea-ſalt), incapable of acting either on the metallic or the ſiliceous body. If the acid was moſt powerful, the compound ſalt might act on the metal, but not at all upon the quartz; and if the alkali was moſt powerful, the compound might act on the quartz, but not at all on the metal. In no caſe, therefore, could it act on both at the ſame time. Fire or heat, if ſufficiently intenſe, is not ſubject to this difficulty, as it could exerciſe its force with equal effect on both bodies.

218. The ſimultaneous conſolidation of the quartz and the metal is indeed ſo highly improbable, that the Neptuniſts rather ſuppoſe, that the ramifications in ſuch ſpecimens as are here alluded to, have been produced by the metal diffuſing itſelf through *rifts* already formed in the ſtone *. But it may be anſwered, that between the channels in which the metal pervades the quartz, and the ordinary cracks or fiſſures in ſtones, there is no reſemblance whatever: That a ſyſtem of hollow tubes, winding through a ſtone, (as the tubes in queſtion, muſt have been, according to this hypotheſis, before they were filled by the metal), is itſelf far more inconceivable than the thing which it is intended to explain;

* Geol. Eſſays, p. 401.

and laſtly, that if the ſtone was perforated by ſuch tubes, it would ſtill be infinite to one that they did not all exactly join, or inoſculate with one another.

219. The compenetration, as it may be called, of two heterogeneous ſubſtances, has here furniſhed a proof of their having been melted by fire. The incluſion of one heterogeneous ſubſtance within another, as happens among the ſpars and druſens, found ſo commonly in mineral veins, often leads to a ſimilar concluſion. Thus, from a ſpecimen of chalcedony, including in it a piece of calcareous ſpar, Dr Hutton has derived a very ingenious and ſatisfactory proof, that theſe two ſubſtances were perfectly ſoft at the ſame time, and mutually affected each other at the moment of their concretion *.

Each of theſe ſubſtances has its peculiar form, which, when left to itſelf, it naturally aſſumes; the ſpar taking the form of rhombic cryſtals, and the chalcedony affecting a mammalated ſtructure, or a ſuperficies compoſed of ſpherical ſegments, contiguous to one another. Now, in the ſpecimen under conſideration, the ſpar is included in the chalcedony, and the peculiar figure of each is impreſſed on the other; the angles and planes of the ſpar are indented into the chalcedony,

* Theory of the Earth, vol. i. p. 93.

dony, and the ſpherical ſegments of the chalcedony are imprinted on the planes of the ſpar. Theſe appearances are conſiſtent with no notion of conſolidation that does not involve in it the ſimultaneous concretion of the whole maſs; and ſuch concretion cannot ariſe from precipitation from a ſolvent, but only from the congelation of a melted body. This argument, it muſt be remarked, is not grounded on a ſolitary ſpecimen, (though if it were it might ſtill be perfectly concluſive), but on a phenomenon of which there are innumerable inſtances.

220. According to this theory, veins were filled by the injection of fluid matter from below; and this account of them, which agrees ſo well with the phenomena already deſcribed, is confirmed by this, that nothing of the ſubſtances which fill the veins is to be found any where at the ſurface. It is not with the veins as with the ſtrata, where, in the looſe ſand on the ſhore, and in the ſhells and corals accumulated at the bottom of the ſea, we perceive the ſame materials of which theſe ſtrata are compoſed. The ſame does not equally hold of metallic veins: "Look, ſays Dr Hutton, into the ſources of our mineral treaſures? Aſk the miner from whence has come the metal in his veins? Not from the earth or air above, nor from the ſtrata which the vein traverſes: theſe do not contain an atom

of the minerals now confidered. There is but one place from whence thefe minerals may have come ; this is the bowels of the earth ; the place of power and expanfion ; the place from whence has proceeded that intenfe heat, by which loofe materials have been confolidated into rocks, as well as that enormous force, by which the regular ftrata have been broken and difplaced *."

221. The above is a very juft and natural reflection ; but if, inftead of interrogating the miner, we confult the Neptunift, we will receive a very different reply. As this philofopher never embarraffes himfelf about preferving a uniformity in the courfe of nature, he will tell us, that though it may be true, that neither the air, the upper part of the earth's furface, nor even the fea, contain at prefent any thing like the materials of the veins, yet the time was when thefe materials were all mingled together in the chaotic mafs, and conftituted one vaft fluid, encompaffing the earth ; from which fluid it was, that the minerals were precipitated and depofited in the clefts and fiffures of the ftrata.

222. It is alleged, in proof of this hypothefis, that mineral veins are found to be lefs rich as they go farther down, whereas they ought to be richer, if they were filled by the projection of melted

* Theory of the Earth, vol. i. p. 130.

melted matter from below. But the fact, that mines are less rich as they descend farther, though it may hold in some instances, is not general, and may therefore be supposed to arise from local causes, such as are, in respect of us, accidental, and beyond the limits to which our theories can be expected to reach. Thus the mines of Mexico and Peru are said to be subject to the preceding rule; but in the mines of Derbyshire and Cornwall, the very contrary is understood to take place. Besides, what we are pleased to call the riches of a mine, are riches relatively to us, and relatively to a distinction which nature does not recognise. The spars and veinstones which are thrown out in the rubbish of our mines, may be as precious in the eyes of nature, as conducive to the great objects of her economy, and are certainly as characteristic of mineral veins, as the ores of silver or gold, to which we attach so great a value. Unless the former are in smaller quantity, or less highly crystallized at great than at small depths, which I believe is not alleged, no conclusion can be drawn from substances, which occupy in general but a small proportion of any vein, and, in their dissemination through it, do not seem to be always guided by the same law.

223. Again, if the veins were filled by deposition from above, we ought to discover in them

such

ſuch horizontal ſtratification as is the effect of depoſition from water, and we ſhould perceive no marks of the materials having been introduced with violence into their place. The Neptuniſts cannot object to the trial of their theory by theſe two facts.

As to the firſt, it is acknowledged, that there is a certain regular diſpoſition of the ſubſtances in mineral veins, as ſtated § 59, but it is one which has hardly any thing in common with the real phenomena of ſtratification. It conſiſts in the diſtribution of the principal ſubſtances in coats parallel to the ſides of the vein, each ſubſtance forming a ſeparate coat. In a vein, for inſtance, containing quartz, fluor, calcareous ſpar, lead, &c. we might expect to find a lining of quartz cryſtals, applied immediately to the walls of the mine, and following exactly the irregularities of their ſurface; next, perhaps, a coat of fluor, then of calcareous ſpar, and laſt of lead-ore in the centre of the vein, the ſame order being obſerved on the oppoſite ſide. Theſe ſucceſſive coats, it is material to remark, are not in planes, but in uneven ſurfaces, of which the inequalities are evidently determined by thoſe of the walls, that is, of the rock which forms the ſides of the vein; neither are they horizontal, but are parallel to the walls, whether theſe be perpendicular or inclined. Here, therefore, there

there is no appearance of the action of that ſtatical law which has directed the arrangement of the other ſtrata, and which tends to make the plane of every ſtratum depoſited by water perpendicular to the direction of gravity. The coating of the veins has therefore been performed under the conduct of ſome other power than that which preſides over aqueous depoſition. If, as the Neptuniſts maintain, the materials in the veins were depoſited by water, in the moſt perfect tranquillity, it is wonderful that we do not find thoſe materials diſpoſed in horizontal layers, acroſs the vein, inſtead of being parallel to its ſides; and it ſeems very unaccountable, that the common ſtrata, depoſited as we are told while the water was in a ſtate of great agitation, have ſo rigorouſly obeyed the laws of hydroſtatics, (§ 38.), and acquired a paralleliſm in the planes of their ſtratification, which approaches ſo often to geometrical preciſion; while the materials of the veins, in circumſtances ſo much more favourable for doing the ſame, have done nearly the reverſe, and taken a poſition, often at right angles to that which hydroſtatical principles require. This is a paradox which the Neptunian ſyſtem has created, and which therefore it is not very likely to reſolve.

224. Mere words ſhould have little power to miſlead, in a ſcience which treats of ſenſible objects,

jects, ſuch as are always eaſily ſubjected to the examination of ſight or of touch; yet there is ſome appearance as if the Neptuniſts were miſled in this, and other inſtances, by the term *ſtratification*. Though an incruſtation on the perpendicular face of a rock has very little affinity to a ſtratum, ſuch as we are accuſtomed to ſee depoſited by water, yet the ſame name being once impoſed on both, mineralogiſts have proceeded to reaſon concerning them, as if they were preciſely the ſame thing, and were both to be aſcribed to the ſame cauſe. Indeed, every perpendicular or highly-inclined bed of ſtone, is inexplicable as an effect of aqueous depoſition, in a ſyſtem, unprovided, as the Neptunian is *, with the means of raiſing up ſuch beds from a horizontal into a vertical poſition. This obſervation may alſo be extended to all caſes of vertical ſtratification. Water cannot directly arrange its depoſites in planes highly inclined, and therefore I have often wondered to ſee the Neptuniſts contending ſo eagerly for the ſtratification of certain rocks, ſuch as granite, which, being vertical, or highly inclined, was much leſs friendly to their ſyſtem than the entire abſence of all ſtratification would have been. I was diſpoſed to admire their candour, when the uſe

* See preceding note.

ufe which they made of the fact convinced me, that I ought only to wonder at their inconfequential reafoning. The Huttonian Theory is, indeed, the only one which poffeffes the means of reconciling the elevation of the ftrata with their horizontal depofition, and which is entitled to confider ftratification, in whatever plane it may be, as originally the work of the ocean. The geologifts who attach themfelves exclufively to the action of water, will never be able to extend the dominion of that element fo far as Dr Hutton has done, by combining it with fire.

225. But, though the Neptunian fyftem were provided with engines, powerful enough to raife up ftrata from a level to a vertical plane, this would avail nothing in the prefent inftance; fince, on no fuppofition, can the incruftations on the perpendicular fides of a vein have ever been horizontal. On no fuppofition, therefore, can thefe incruftations be received as a proof of aqueous depofition: it may indeed be certainly inferred from them, that the matter which they confift of was fluid at the time of their formation; but the abfence of all appearance of a horizontal difpofition, in any part of the vein, amounts nearly to a demonftration, that this fluidity did not proceed from folution in a menftruum. We muft therefore conceive the coats to have been formed during the refrigeration of the

melted

melted matter injected from the mineral regions into the clefts and fiffures of the ftrata (§ 59.).

226. Mineral veins, particularly at their interfections with one another, contain abundant marks of the moft violent and repeated difturbance, (§ 56). Not to mention that they owe their firft formation to the fracture and difplacing of rocks already confolidated, it appears, that they have originated at very different periods, and that the birth of each has been accompanied with convulfions, which fhook the foundations of the earth. In Cornwall, for inftance, the principal veins, and thofe which they diftinguifh particularly by the name of *Lodes*, have nearly the fame direction with the ftrata or vertical fchiftus, extending from about E. N. E. to W. S. W. Thefe, however, are often interfected nearly at right angles by other mineral veins, called *Crofs Courfes*, and this hardly ever happens without the latter moving, or, as it is called, *heaving* the former out of their direction. This plainly indicates, that the crofs courfes are of later origin than the others, and that their formation was accompanied with fuch a force, as muft, in many inftances, have moved the whole body of rock which conftitutes the promontory of Cornwall, and probably much more, for feveral yards, in a horizontal direction. Sometimes, alfo, both the longitudinal and

and the croſs vein are forced out of their place by a third. Theſe diſturbances ariſe not only from mineral veins, but from veins of porphyry and granite, the production of which has been attended with no leſs violence than of the others.

227. What is here ſaid of Cornwall, is the hiſtory, in ſome degree, of all mineral countries whatever. The great horizontal *tranſlation* which has thus accompanied the formation of veins; the movement impreſſed on ſuch vaſt bodies of rock, and the frequent renewal of theſe immenſe convulſions; are not to be explained by the mild and tranquil dominion of the watery element. They require the utmoſt power that is known any where to exiſt, and were it not for the admonitions of the volcano and the earthquake, we might doubt if even ſubterraneous heat itſelf poſſeſſed an energy adequate to theſe aſtoniſhing effects.

228. From the *heaving* of one vein by another, it is evident, that there was a force of protruſion in the direction of one of them, that acted at the time of its formation. This force cannot be accounted for on the ſuppoſition that veins were produced by the mere ſhrinking of the ſtrata; for the rocks could not, in that caſe, have been rent aſunder, and impelled forward at the ſame time. It appears moſt likely, that fiſ-

ſures

ſures in the ſtrata were made, at leaſt in many inſtances, and the matter poured into them, nearly at the ſame time, both being effects of the ſame cauſe, the expanſive force of ſubterraneous heat.

229. It is remarked, at § 56., that the ſhifting of the ſtrata is beſt obſerved where the veins make a tranſverſe ſection of beds of rock, conſiderably inclined to the horizon. It is alſo true, that in ſome caſes the near approach of the ſtrata to the level, may make the ſhifts produced by the veins very eaſy to be diſcovered. Thus in Derbyſhire, where the mineral veins are in ſecondary ſtrata, nearly horizontal, there is almoſt no inſtance in which the correſponding ſtrata are not obſerved to be on different levels, on the oppoſite ſides of the ſame vein.

230. The fact deſcribed by De Luc, and referred to at § 55., may, for what we know of it, admit of being explained in two ways. The great wedge of rock which appears to be inſulated between two branches of the ſame vein, may either be a maſs that has been broken off, and ſuſtained by the melted matter that flowed all around it; or, it may be a maſs of rock contained between two veins that are in reality diſtinct, and of different formation. Whether this laſt ſuppoſition is the truth, would probably be evident from a careful examination of both

parts

parts of the vein; as ſome difference of character cannot fail to be the conſequence of different formation. If no ſuch difference is obſerved, the two branches muſt be ſuppoſed to belong to the ſame vein, and the only probable explanation of the inſulation of ſo large a maſs of rock will be by the firſt-mentioned ſuppoſition. This fact, therefore, notwithſtanding the great attention M. De Luc has beſtowed on it, ſtill requires further examination, before it can be decided whether it inclines to the Huttonian Theory, as on the firſt ſuppoſition, or is, as on the latter hypotheſis, equally balanced between it and the *Wernerian*.

231. Whatever be the caſe with this fact, the general one of pieces of rock being found inſulated in veins, is certainly favourable to the notion of an injected and ponderous fluid having originally ſuſtained them. Where, as happens in ſome inſtances, the ſtones contained in the veins have no affinity to any of the rocks above, they cannot be ſuppoſed to have come any how but from below, and to have been carried up by the matter of the vein. The inſtance from the ſlip at the Huddersfield Canal has been already mentioned.

232. The preceding obſervations have been principally directed againſt that theory of veins which ſuppoſes them to have been filled by de-

 poſition

pofition from water. There is another theory maintained by fome of the Neptunifts, that the metals in veins were introduced there by infiltration *. This opinion is fufficiently refuted by the fact, that rarely any metallic ore is found out of the vein, or in the rock on either fide of it, and leaft of all where the vein is richeft. This is inconfiftent with the notion of the ore being carried into the vein by water percolating through the adjacent rocks, unlefs fome fatisfactory reafon is affigned, which determined the water to leave the ore in the vein and no where elfe. Befides, this hypothefis does not account for the formation of the fpars and veinftones which fill the vein, and which appear clearly to have been brought there at the fame time with the ore, and no doubt by the fame caufe.

233. The veins, properly fo called, are indefinitely extended; but there are alfo thin plates of fpar, and of cryftals of different kinds, often found included in rocks, and fhut in on all fides, to which the name of veins is commonly applied. Thefe laft ought certainly to be diftinguifhed from the former, and may not improperly be called *Plate Veins* or *Lenticular Veins*, the plate or cake of fpar of which they confift having very often the form of a lens, though, as

* Geol. Effays, p. 401.

as may be ſuppoſed, conſiderably irregular. Either of theſe terms being derived entirely from external characters, has the advantage of involving nothing theoretical.

The lenticular veins are certainly not formed like the uſual mineral veins, by injection, ſince they are ſhut in, on all ſides, by the ſolid rock. When they are found, therefore, in ſtratified rocks, ſuch as have not themſelves been melted, we muſt conceive them to be compoſed of materials more fuſible than the ſurrounding rock, ſo that they have been brought into fuſion by a degree of heat which the reſt of the rock was able to reſiſt, and, on cooling, have aſſumed a ſparry ſtructure. When they are found in rocks, of which the whole has been fluid, they muſt be conſidered as component parts of that maſs, which, by an elective attraction, have united with one another, and ſeparated themſelves from the ſubſtances to which they had leſs affinity.

The veins of this kind ſeem to be connected with thoſe called in Derbyſhire *Pipe Veins*, in which the ores of metals are ſometimes found. The pipe veins, indeed, are not in all caſes completely inſulated, but ſometimes communicate with the veins properly called mineral. I am too little acquainted, however, with their natural hiſtory, to be able to ſay with certainty to

 which

which of the two ſpecies they ought to be referred.

Note xiv. § 75.

On Whinſtone.

234. To the facts and reaſonings given above, I ſhall, in this note, add a few remarks, tending to ſhew, that whinſtone is not of volcanic, nor of aqueous, but certainly of igneous origin.

It is aſſerted (§ 62.), that carbonat of lime and zeolite are often contained in whinſtone, but never in lava, and that this circumſtance may ſometimes ſerve to diſtinguiſh theſe ſtones from one another. With reſpect to carbonat of lime, in particular, it ſeems evident, that this ſubſtance cannot enter into the original compoſition of any lava, becauſe the ſame heat which melted the lava, would, where there was no greater preſſure than the weight of the atmoſphere, expel the carbonic acid and produce quicklime. Notwithſtanding this, rocks, containing carbonat of lime, have often been conſidered as lavas, into the pores and cavities of which, calcareous matter having been carried by the infiltration of water, had cryſtallized into ſpar. Thus Spallanzani, in in his account of the Euganean Hills, in Lombardy, defcribes fome of the rocks as a ounding at their furface, and even in th interior, with air-bubbles of various fizes, from such as are hardly perceptible, to fome that are half an inch in diameter ; and which, he fays, are all of an oval figure, with their longeft diameters in the fame direction. This he confiders as a proof that the rock is a genuine lava ; for the air-bubbles prove the ftone to have had its fluidity from fire ; and by their elongation in the fame direction they prove, that the mafs when fluid was alfo in motion. Spallanzani adds, that *many of thefe cavities are filled with cryftals of the carbonat of lime, an effect of the infiltration of water* *.

235. Though the argument here advanced for the igneous origin of the rock may be admitted as conclufive, the introduction of calcareous fpar into it by infiltration muft ftill be queftioned. Lava, except in a ftate of decay or decompofition, is not readily penetrated by water ; and, if it were, the filling of cavities with fpar, by means of the water percolating through them, would ftill be fubject to many difficulties, (§ 12.). Befides, whinftone rocks are frequently found

 fo

* Voyages dans les deux Siciles, tom. iii. p. 157. Edit. de Faujas de St Fond.

ſo full of calcareous ſpar, or of zeolite, that they would become porous to ſuch a degree, if the cavities filled with theſe latter ſubſtances were all empty, that they could hardly ſuſtain their own weight, and much leſs that of the great maſſes of rock incumbent on them. In ſuch caſes, it is certain, that the cryſtallized ſubſtances were part of the original compoſition of the rock. The truth is, that the infiltration of the water is a mere gratuitous aſſumption, introduced for the purpoſe of explaining the exiſtence of carbonated lime in a ſtone which had endured the action of intenſe heat; and this aſſumption ought of courſe to be rejected, if the phenomenon can be explained by a theory, that is in other reſpects conformable to nature. The ſpar, then, may be conſidered as a proof, that the rocks in queſtion are to be numbered with thoſe unerupted lavas which have flowed deep in the bowels of the earth, and under a great compreſſing force. This is the more probable, that the Euganean Hills, like ſome whinſtone hills in our own country, have, in certain places, a covering of ſlaty and calcareous ſtrata incumbent on them, even at their ſummits *, ſo that the torrent of melted ſtone, of which they are admitted to conſiſt, cannot have flowed from the mouth of a volcano. I do not

* Phil. Tranſ. 1775, p. 34.

not mean to ſay, that there are among theſe hills no veſtiges of volcanic exploſion. I am very far from having *data* ſufficient for drawing this concluſion ; but I believe it may be ſafely affirmed, that the bulk of them is no more compoſed of volcanic lava, than the baſaltes of Staffa, or of the Giant's Cauſeway.

236. But, beſides the evidence deduced from calcareous ſpar and zeolite, againſt the rocks containing them being real lava, there are other marks, even leſs equivocal perhaps, that diſtinguiſh the lavas which we ſuppoſe to have flowed in the mineral regions, from thoſe which have actually flowed on the ſurface. Theſe are what we collect from the diſpoſition, the organization, or, as we may ſay, the phyſical geography of whinſtone countries, unlike, in ſo many reſpects, to that of volcanic countries. The ſhape of whinſtone hills ; their large flat terraces, riſing one above another ; their perpendicular faces, and the correſpondence of their heights even at conſiderable diſtances ; have nothing ſimilar to them in the irregular torrents of volcanic lavas. The phenomena of the former are alſo on a ſcale of magnitude very far exceeding the latter, and clearly indicate, that though both have been produced by fire, it has been by fire in very different circumſtances, and regulated by very different laws. The ſtructure of the two kinds of

rock agrees, in many respects, and so does their chemical analysis; but their disposition and arrangement are so dissimilar, that they cannot be supposed to be of the same formation.

237. This argument, I believe, was first stated by Mr STRANGE, in a letter to Sir JOHN PRINGLE, published in the 65th volume of the *Philosophical Transactions* *. That intelligent observer, after visiting the countries in Europe most remarkable either for burning, or for what are accounted, extinguished volcanoes, and examining them with a very discriminating eye, remained convinced, that there are two distinct species of rock, which both owe their origin to fire; but to fire acting in circumstances and situations extremely different. The first is the common volcànic lava; the other, to which he gives the name of a basaltine rock, comprehends such rocks as the Giant's Causeway, the basaltes of the Vivarais, of the Euganean Hills, &c. and differs in nothing from that which is called here by the name of whinstone. Mr Strange conceived, that the one of these kinds of stone could, no more than the other, be accounted the work of aqueous deposition, but was led to the distinction just mentioned, by observing the organization and

* Account of Two Giants Causeways in the Venetian State, &c. by John Strange, Esq; Phil. Transf. vol. lxv. (1775.) p. 5, &c.

and arrangement in the rocks of the latter kind, and comparing them with the diſorder and ruin that every where mark the footſteps of volcanic fire. He does not pretend to determine the nature of the fire to which the baſaltine rocks owe their formation, nor the circumſtances in which it has acted: he is ſatisfied with the negative concluſion, that it is not volcanic; and his paper affords a ſpecimen of what is perhaps rare in any of the ſciences, and certainly moſt rare of all in geology, viz. a philoſophic induction carried juſt as far as the facts will bear it out, and not a ſingle ſtep beyond that point.

238. Several other hints contained in this paper are highly deſerving of notice; for we not only find in it the notion of a formation of baſaltic rocks, igneous though not volcanic, but alſo that of their ſimultaneous cryſtallization *, together with the ſuggeſtion, that granite and baſalt are of the ſame origin †. Theſe opinions had not, I believe, occurred at that time to any mineralogiſt except Dr Hutton, nor had they been communicated by him to any but a few of his moſt intimate friends; ſo that Mr Strange has without doubt all the merit of a firſt diſcoverer. Indeed, without the knowledge of the

* Phil. Tranſ. *ubi ſupra*, p. 17.

† *Ibid.* p. 36. and 37.

the principle of compreſſion, ſuch as it is laid down by Dr Hutton, it was hardly poſſible for him to proceed further than he has done. He remarked the *unburnt* limeſtone that lies on the tops of ſome of the Euganean baſaltes, and ſeems to have been aware of the great difficulty, which it was reſerved for the Huttonian Theory to overcome. His letter contains alſo ſome excellent general remarks on the rocks of the Vivarais and Velay, which he had viſited, before FAUJAS DE ST FOND had publiſhed his curious and elaborate deſcription of theſe countries.

239. The cauſe of the peculiar ſtructure which has juſt been obſerved to diſtinguiſh whinſtone from volcanic countries, is eaſily aſſigned in the Huttonian Theory. According to that theory, the whinſtone rocks were formed, in the bowels of the earth, of melted matter poured into the rents and openings of the ſtrata. They were caſt, therefore, in thoſe openings, as in a mould; and received the impreſſion and character of the rocks by which they were ſurrounded. Hence the tabular maſſes of whinſtone, which when ſoft have been interpoſed between ſtrata, and compreſſed by their weight, ſo as almoſt to have themſelves acquired the appearance of ſtratification. Hence the perpendicular faces of the ſame rocks, produced by their being abutted when

yet

yet ſoft, againſt the abrupt ſides of the ſtrata. The rocks which formed thoſe moulds have, in many caſes, entirely diſappeared ; in others, a part ſtill remains, ſurrounding, or even covering, the baſaltes, as in the Euganean Hills, in thoſe of the Val di Noto in Sicily, the rocks near Liſbon *, and in different parts of Great Britain.

Above all, the veins of whinſtone which interſect the ſtrata, are the completeſt proofs of the theory here given of theſe rocks, and the moſt inconſiſtent, in all reſpects, with the hypotheſis of their volcanic origin.

240. If theſe *criteria* are applied to what are called extinguiſhed volcanoes, I have no doubt that many which have been reckoned of that number, will be found to derive their origin more directly from the fire of the mineral regions. The baſaltic rocks of the Vivarais, I am well perſuaded, belong to this claſs ; and I conclude that they do ſo, not only from the account of them given by Mr Strange, but from the deſcription of Faujas himſelf, who, though under the influence of the oppoſite theory, ſeems very fair and accurate in his deſcription of phenomena. The moſt unequivocal mark of real whinſtone rock, and of a formation in the ſtrict-

eſt

* Recherches ſur les Volcains Eteints du Vivarais ; Lettre de Dolomieu, p. 443.

eſt ſenſe mineral, is where veins of that kind of rock interſect the ſtrata. Now, in a letter to Buffon, on the ſtreams of lava found in the interior of certain calcareous rocks in the lower Vivarais, Faujas deſcribes what can be accounted nothing elſe but a vein or dike of whinſtone, accompanied with ſeveral of its moſt remarkable and characteriſtic appearances: "Figurez-vous un courant de lave, de la nature du baſalte noir, dur et compacte, qui a percé à travers les maſſes calcaires, et s'eſt fait jour dans quelques parties, paroiſſant et diſparoiſſant alternativement: Cette coulée de matière volcanique s'enfonce ſous une partie de la ville, bâtie ſur le rocher; elle reparoit dans la cave d'un maréchal, ſe cache et ſe montre encore de temps en temps en deſcendant dans le vallon, &c. Ce qu'il y a d'admirable, c'eſt que la lave forme deux branches bien extraordinaires, dont l'une s'éleve ſur la crête du rocher, tandis que l'autre coupe horizontalement de grands bancs calcaires eſcarpés, qui ſont à découvert, et bordent le chemin.

"Quels efforts n'a-t-il pas fallu pour forcer cette lave ſe prendre une telle direction, et ſe percer cette ſuite de rochers calcaires? Si cette longue coulée de lave avoit eu 200 ou 300 toiſes de largeur, je ne ſerois pas ſurpris qu'un torrent de matiere en fuſion de ce volume eut pu produire des effets extraordinaires et violens;

mais

mais figurez-vous, Monſieur, que dans les endroits les plus larges, elle n'a tout-au-plus qu'environ 12 *ou* 15 *pieds ; elle n'en a que* 3 *ou* 4 *dans certaines parties* *."

This narrow ſtream is to be traced acroſs the ſtrata for more than a league and a half; and the whole appeared to Faujas ſo marvellous, that he ſays he almoſt doubted the teſtimony of his ſenſes. He would have done much better, however, to have doubted the concluſions of his theory; for it was by them that the phenomena before him were rendered ſo myſterious and incredible. While he continued to regard what is deſcribed above as a ſtream of melted lava, which had deſcended from the top of one mountain, and climbed up the ſides of the oppoſite, like water in a conduit pipe, piercing occaſionally through vaſt bodies of ſolid rock, it is no wonder that he conſidered as marvellous what is indeed phyſically impoſſible. Had his belief in the volcanic theory permitted him to ſee in all this, not a ſuperficial current, but one of indefinite depth, he would have beheld the object diveſted, not of what was curious and intereſting, but of what was incredible or abſurd, and reduced to the ſame claſs of things with mineral veins. That it belongs really to this claſs, and is no more than a vein or dike of

* Volcains Eteints du Vivarais, p. 328, &c.

of whinſtone, interſecting the ſtrata to an unknown depth, and moſt probably, like other veins, communicating with the mineral regions, cannot be doubted by any one who has ſtudied the ſubject of baſaltine rocks, through any other medium than the volcanic theory. The ramifications which run from it into the calcareous rock, contrived, Faujas ſays, juſt as if on purpoſe to perplex mineralogiſts, is one of the well-known and characteriſtic appearances of baſaltic veins.

241. It can hardly be doubted, that the lava deſcribed by the ſame author as heaving up a maſs of granite*, and including pieces of it, is a rock of real whinſtone. The ſame may be ſaid of many others; and, though I pretend not to affirm that there is nothing volcanic in the Vivarais, I muſt ſay, that nothing decidedly volcanic appears in the deſcription of that country, but many things that are certainly of a very different origin.

In the preſent ſtate of geological ſcience, a ſkilful mineralogiſt could hardly employ himſelf better, than in traverſing thoſe ambiguous countries, where ſo much has been aſcribed to the ancient operation of volcanic fire, and marking out what belongs either clearly to the erupted

* Volcains Eteints du Vivarais, fol. p. 365, &c.

ed or unerupted lavas, and what parts are of doubtful formation, containing no mark by which they may be referred to the one of thefe any more than to the other. Such a work would contribute very materially to illuftrate the natural hiftory of the earth.

242. One of the moft ingenious attempts to fupport the volcanic theory, is the fyftem of *fubmarine volcanoes*, imagined by the celebrated mineralogift DOLOMIEU. The phenomenon that led to this hypothefis, was what he had obferved in the hills near Lifbon, and ftill more remarkably in thofe of the Val di Noto in Sicily, where the bafaltine rocks had regular ftrata incumbent on them, and in fome cafes interpofed or alternated with them *. It feemed from this evident, that the ftrata were of later formation than the ftone on which they refted; and as they muft, on every fuppofition, be held to be depofited by water, it was concluded, that the lava which they covered had been thrown out by volcanoes at the bottom of the fea; that the ftrata had afterwards been depofited on this lava; and that, in fome cafes, there had been frequent

* Memoire de Deodate de Dolomieu, fur les Volcains Eteints du Val di Noto, en Sicile. Journal de Phyf. tom. xxv. (1784. Septembre.) p. 191.

quent alternations of theſe eruptions and depoſitions *.

243. Though this hypotheſis does certainly deliver the ſyſtem of the Volcaniſts from one great difficulty, it is itſelf liable to inſurmountable objections. I ſhall juſt mention ſome of the principal.

1. The regular and equidiſtant ſtrata that we often ſee covering the tops of whinſtone or baſaltic rocks, could not have been depoſited in the oblique and very much inclined poſition which they now occupy.

This is remarkable in the ſtrata which cover the baſaltic rock of Saliſbury *Craig*, near Edinburgh, at its northern extremity. The ſtrata are very regular, and muſt have been depoſited in a plane nearly horizontal; yet the ſurface of the baſaltes on which they now reſt is very much inclined, dipping rapidly to the north-eaſt. The neceſſity of a horizontal depoſition in ſtrata, which, though not now horizontal, have their planes

* Near Vizini, in the Val di Noto, Dolomieu tells us, that he counted eleven beds, alternately calcareous and volcanic, in the perpendicular face of a hill, which at a diſtance appeared like a piece of cloth, ſtriped black and white; *ubi ſupra*. In another inſtance he ſaw more than twenty of theſe alternations. He has ſince made ſimilar obſervations in the Vincentine and in Tirol. Journal de Phyſ. tom. xxxvii. (1790), partie 2. p. 200.

planes nearly parallel to one another, has been proved at § 38.

2. If there is any truth in the principles eſtabliſhed above, even the ſtrata themſelves have not been conſolidated without the action of fire. By Dolomieu's ſyſtem, therefore, the conſolidation of the ſtrata which cover the baſaltes is not accounted for.

3. There are no means furniſhed by the hypotheſis of ſubmarine volcanoes for bringing the baſalt, and the ſtrata which cover it, above the level of the ſea. If it is ſaid that the waters of the ſea have been drained off, the objections are all incurred that have been ſtated at § 37 *. If it is ſaid, that the rocks themſelves have been elevated by a force, impelling them upwards, we ſay, that the exiſtence of ſuch a force, when admitted, furniſhes another means of explaining the whole phenomenon, namely, that of the injection of melted matter among the ſtrata, the ſame that is uſed in the Huttonian Theory.

4. The phenomena of balſaltic veins are not in the leaſt explained by the hypotheſis of ſubmarine volcanoes. That hypotheſis, then, even if the foregoing objections were removed, does

 not

* Dolomieu adopts this ſuppoſition; he thinks, that the ſurface of the ſea muſt have been formerly 500 or 600 toiſes above its preſent level. *Ibid.* p. 196.

not ſerve to explain all the facts reſpecting the rocks of this genus, and wants, of conſequence, one of the moſt important characters of a true theory. It muſt be allowed, however, that it makes a conſiderable approach to ſuch a theory, and that the ſubmarine volcanoes of Dolomieu, have an affinity to the unerupted lavas of Dr Hutton.

244. Though in theſe remarks I have endeavoured to expoſe the errors of the volcanic ſyſtem, I cannot but conſider that ſyſtem as coming infinitely nearer to the truth than the Neptunian. It has the merit of diſtinguiſhing an order of rocks, which bears no mark of aqueous formation, and in which the cryſtallized, ſparry, or lava-like ſtructure, beſpeaks their primeval fluidity, and refers their origin to fire. The Neptunian ſyſtem, on the other hand, ſtrives to confound the moſt marked diſtinction in the mineral kingdom, and to explain the formation, both of the ſtratified and unſtratified rocks, by the operation of the ſame element. Though chargeable with this inconſiſtency, it has become the prevailing ſyſtem of geology; and the arguments which ſupport it are therefore entitled to attention.

245. It will no doubt be thought ſingular, that the ſame mineralogiſt, whom we have juſt ſeen exerting his ingenuity in defence of the volcanic

volcanic ſyſtem, ſhould now appear equally ſtrenuous in defence of the Neptunian. Though Dolomieu contends for the volcanic origin of ſome baſaltic rocks, he does not admit that all baſaltes is volcanic, nor even all of igneous formation. Thus he ſtates, that he had examined at Rome ſome of the moſt ancient monuments of art, executed in baſaltes, brought from Upper Egypt, and that he could diſcover no mark of the action of fire in any of them *. On the contrary, he found that ſome of them conſiſted of green baſaltes, which changes its colour to a bronze, when expoſed even to a moderate heat, and which therefore, he argues, can never have endured any ſtrong action of fire.

The anſwer to this argument is very plain, if we admit the effects aſcribed by Dr Hutton to the compreſſion which neceſſarily takes place in the mineral regions. If indeed the heat in thoſe regions reſembled exactly that of our fires at the ſurface, it would not be eaſy to deny the above concluſion, which therefore certainly holds good againſt the volcanic origin of the Egyptian baſaltes. But there is no reaſon why, under ſtrong compreſſion, the colouring matter

 of

* Journal de Phyſique, tome xxxvii. (1790.) partie 2. p. 193.

of theſe ſtones might not be fixed, and indeſtructible by heat, though it can be eaſily volatilized or conſumed when ſuch compreſſion is removed. This argument then is againſt the volcanic ; but not againſt what has been called the *Plutonic* formation of baſaltes.

246. As to the other marks of fire which Dolomieu ſought for and did not find in the abovementioned ſtones, we are not exactly informed in what they conſiſted. If the cryſtallized or ſpathoſe texture that belongs to this deſcription of ſtones was wanting, the ſpecimens were not to be conſidered as of the real baſaltic or whinſtone genus, whatever their name or hiſtory may ſeem to indicate. If they did poſſeſs that texture, they had the only mark of an igneous origin that could be expected, ſuppoſing that origin to have been in the bowels of the earth. No part, therefore, of the obſervations of this ingenious mineralogiſt, can be conſidered as inconſiſtent with the theory of baſaltic rocks which has been laid down above.

247. Bergman had before reaſoned on this ſubject preciſely in the ſame manner, but from better data, as the ſtones from which he derived his argument were in their native place: " Trap," ſays that ingenious author, (that is, whinſtone), " is found in the ſtratified mountains of Weſt Gothland, in a way that deſerves

to

to be defcribed. The lower ftratum, which is feveral Swedifh miles in circuit, ($10\frac{1}{2}$ of thefe miles make a degree), is an arenaceous ftone, horizontal, refting on granite, and having its particles agglutinated by clay. The ftratum above this is calcareous, full of the petrifactions of marine animals, and above this is the trap. Thefe three kinds of rock compofe the greater part of the mountains juft mentioned, though there are fome other beds, particularly very thin beds of marl and of clay, which feparate the middle ftratum, both from that which is under it and over it, and are frequently fo penetrated with bitumen that they burn in the fire. This fchiftus is black; when burnt it becomes red, and afterwards, when wafhed with water, affords alum. How can it be fuppofed," he adds, "that the trap has ever been violently heated, while the fchiftus on which it is incumbent retains its blacknefs, which however it lofes by the action even of a very weak fire *."

The anfwer to this argument is already given. The reafoning, as in the former inftance, is conclufive only againft the action of volcanic fire, or fire at the furface; but not againft the action of heat deep in the bowels of the earth, and un-

* Bergman de Productis Volcaniis Opufcula, tom. iii. p. 214, &c.

der the preſſure of the ſuperincumbent ocean. In ſuch a ſituation, the bituminous ſchiſtus might be in contact with the melted baſalt, and yet there might be no evaporation of the volatile, nor combuſtion of the inflammable parts. It does not, however, always happen, that the bituminous ſubſtances, or ſubſtances alterable by fire, which are found in contact with baſaltes, are without any mark of having endured the operation of fire. Inſtances in which ſuch operation is apparent are given above, § 30.; and more will be added in the concluſion of this note.

248. The ſame mineralogiſt founds another argument for the aqueous formation of whin or trap, on the exiſtence of that ſtone in the form of veins, included in primeval rocks: "Invenitur hoc ſaxum (trap) in Suecia pluribus locis, ſæpeque in montibus primævis, anguſtas implens venas, adeo ſubtilis ſtructuræ, ut particulæ ſint impalpabiles, et, dum niger eſt, genuinum efficit lapidem Lydium. In hiſce montibus, nulla adſunt ignis ſubterranei veſtigia *."

The phenomenon here deſcribed, namely, a vein of compact whinſtone traverſing a primary rock, is, without doubt, as incapable of being explained by the operation of a volcano, as it is by

* Opuſcula, ubi ſupra.

by that of aqueous depofition. It is, however, a moft complete proof of the original foftnefs of the fubftance of which the veins confift, and affords one of the ftrongeft poffible arguments for fuch an operation of fire as is fuppofed in the prefent theory. The main arguments, therefore, which have been propofed as fubverfive of the igneous origin of bafaltes, are only fubverfive of their formation by one modification of fire, viz. of fire acting near the furface; and thus the weapons which directly pierce the armour of the Volcanift, and inflict a mortal wound, are eafily turned afide by the fuperior temper of the *Plutonic* mail.

249. An argument founded on facts very fimilar to fome of the preceding, and leading to the fame conclufion, is employed by the mineralogift to whom the Neptunian fyftem owes its chief fupport. Werner, in his obfervations on volcanic rocks and on bafaltes, has refted his proof of the aqueous formation of the latter, on their interpofition between beds of ftone in mountains regularly ftratified, and obvioufly formed by water. He defcribes an inftance of this in the bafaltic hill of *Scheibenberg;* and the facts, though moft of them are not uncommon, are highly deferving of attention. Near the top of this hill, and above the bafaltic rock which compofes the body of it, he tells us that

there was a ſand-pit; a circumſtance which he appears to conſider as not a little ſingular. It was, however, at the bottom of the hill, that he met with the appearances which chiefly attracted his notice: "Firſt," ſays he, "or loweſt, was a thick bank of quartzy ſand, above that a bed of clay, then a bed of the argillaceous ſtone called wacken, and upon this laſt reſted the baſaltes." "When I ſaw," adds he, "the three firſt beds running almoſt horizontally under the baſaltes, and forming its baſe; the ſand becoming finer above, then argillaceous, and at laſt changing into real clay, as the argil was converted into wacken in the ſuperior part; and, laſtly, the wacken into baſaltes: in a word, when I found a perfect tranſition from pure ſand to argillaceous ſand, from the latter to a ſandy clay, and from this ſandy clay, through many gradations, to a fat clay, to wacke, and at laſt baſaltes, I was irreſiſtibly led to conclude, that the baſaltes, the wacke, the clay, and the ſand, are all of one and the ſame formation; and that they are all the effect of a chemical precipitation during one and the ſame ſubmerſion of this country *."

Firſt,

* "Combien je fus ſurpris de voir en arrivant au fond, un épais *banc de ſable quartzeux*, puis au-deſſus une *couche d'argile*, enfin une couche de la pierre argileuſe nommée *Wacke*, et ſur celle-ci repoſer le *baſalte*. Quand je

First, as to the ſand on the top of this baſaltic hill, it is moſt probably the remains of certain ſandſtone ſtrata that originally covered the baſaltic part, but are now worn away. We are therefore to conſider this as an inſtance of a baſaltic rock, interpoſed between ſtrata that are undoubtedly of marine origin. In this, however, there is nothing inconſiſtent with Dr Hutton's theory of baſaltes; on the contrary, it is one

---

je vis les trois premières couches s'enfoncer *preſqu' horizontalement ſous le baſalte*, et former ainſi ſa *baſe;* le ſable devenir plus fin au-deſſus, puis argileux, et ſe changer enfin en vraie argile, comme l'argile ſe convertiſſoit en wacke dans ſa partie ſupérieure; et finalement la wacke en baſalte: en un mot, de trouver ici une *tranſition parfaite* du *ſable pur* au *ſable argileux*, de celui-ci a *l'argile ſabloneuſe*, et de *l'argile ſabloneuſe*, par pluſieurs gradations, à l'*argile graſſe*, à la *wacke*, et enfin au *baſalte*.

" A cette vue, je fus ſur-le-champ et irréſiſtiblement entrainé à penſer, (comme l'auroit été ſans doute tout connoiſſeur impartial frappé des conſéquences de ce phenomène); je ſus, dis je, irréſiſtiblement entrainé aux idées ſuivantes: Ce *baſalte*, cette *wacke*, cette *argile*, et ce *ſable*, *ſont d'une ſeule et même formation;* ils ſont tous l'effet d'une *precipitation par voie humide* dans une ſeule et même ſubmerſion de cette contrée; les eaux qui la couvroient alors tranſportoient d'abord le *ſable*, puis depoſoient l'*argile*, et changoient peu-à-peu leur précipitation en *wacke*, et enfin en vraie *baſalte*." — Journal de Phyſique, tome xxxviii. (1791), partie 1. p. 415.

one of the principal facts on which that theory is founded. It has indeed been argued by ſome mineralogiſts, that bodies thus contiguous muſt owe their origin to the ſame element, and that a mineral ſubſtance cannot be of more recent formation than that which lies above it. But the maxim, that a foſſil muſt have the ſame origin with thoſe that ſurround it, does not hold, unleſs they have a certain ſimilarity of ſtructure. It is, for inſtance, the want of this ſimilarity, that authoriſes us to aſſign different periods of formation to mineral veins, and to the rocks in which they are included.

In a ſucceſſion of ſtrata, no one can doubt, that the loweſt were the firſt formed, and the others in the order in which they lie ; but, when between two ſtrata of ſandſtone or of limeſtone we find an intermediate rock, ſo different as to reſemble lava, and to have nothing ſchiſtoſe or ſtratified in its compoſition, the ſame inſtrument cannot be ſuppoſed to have been employed in the formation of both ; nor is there any reaſon why we may not ſuppoſe, that the intermediate body was interpoſed between the other two, by ſome action ſubſequent to their formation. It was thus that Dolomieu concluded, when he ſaw a lava-like ſtone interpoſed between calcareous ſtrata in the Val di Noto, that, though contiguous,

contiguous, theſe two rocks could not poſſibly be of the ſame formation ; and thus far it is certain, that every unprejudiced obſerver muſt agree with him.

250. But the circumſtance on which Mr Werner ſeems to lay the greateſt ſtreſs, is the gradual tranſition from the ſand to the baſalt, through the intermediate ſteps of clay and wacken ; this gradual tranſition he conſiders as a direct proof, that they are all of the ſame formation.

A gradual tranſition of one body into another, can only be ſaid to take place, when it is impoſſible to define their common boundary, or to determine the line where the one begins and the other ends. Now, if this be the proper notion of gradual tranſition, I muſt ſay, that after much careful examination, I have never ſeen an inſtance, in which ſuch a tranſition takes place between whinſtone and the contiguous ſtrata. The *line* of ſeparation, though in ſome places leſs evident than in others, has, on the whole, been marked out with great preciſion ; and, though the ſtones have been firmly united, or, as one may ſay, welded one upon another, yet, when a freſh fracture was obtained, the ſtratified and unſtratified parts have rarely failed to be diſtinguiſhed. The freſh fracture is indeed often neceſſary, for many ſpecies of whinſtone

ſtone get by decompoſition a granulated texture at the ſurface, ſo as hardly to be diſtinguiſhed from real ſandſtone.

Some of the kinds of primary ſchiſtus alſo, particularly the argillaceous, when much indurated, have in their ſtructure a conſiderable reſemblance to whinſtone ; they are ſlightly granular, or laminated, and have a tendency to a ſparry texture. Where it happens that this ſort of ſchiſtus and whinſtone are contiguous, it is natural to expect, that their common boundary will be traced with difficulty, and in many parts will be quite uncertain. Still, however, if a careful examination is made ; if the effects of accidental cauſes are removed ; and, above all, if the more ambiguous inſtances are compared with the more deciſive, and interpreted by them, though ſingle ſpecimens may be doubtful, we will hardly ever find that any uncertainty remains with reſpect to entire rocks.

251. This general fact, which I ſtate on much better authority than that of my own obſervations, viz. on thoſe of Dr Hutton, is not given as abſolutely without exception. The theory of whinſtone which has been laid down here, leads us indeed to look for ſome ſuch exceptions. It is certain, that the baſis of whinſtone, or the material out of which it is prepared by the action

tion of ſubterraneous heat, is clay in ſome ſtate or other, and probably in that of argillaceous ſchiſtus. It follows, of conſequence, that argillaceous ſchiſtus may by heat be converted into whinſtone. When, therefore, melted whinſtone has been poured over a rock of ſuch ſchiſtus, it may, by its heat, have converted a part of that rock into a ſtone ſimilar to itſelf; and thus may now ſeem to be united, by an inſenſible gradation, with the ſtratum on which it is incumbent; and phenomena of this kind may be expected to have really happened, though but rarely, as a particular combination of circumſtances ſeems neceſſary to produce them. Hence it is evident, that ſtones may graduate into one another, without being of the ſame formation; and that it is fallacious to conclude, from the inſenſible tranſition of one kind of rock into another, without any other circumſtance of affinity, that they have both the ſame origin.

I am diſpoſed, therefore, to make ſome limitation to what is ſaid in § 72, where I have expreſſed an abſolute incredulity as to ſuch tranſitions as are here referred to. The great ſkill and experience of the mineralogiſt who has deſcribed the ſtrata at Scheibenberg, do not allow us to doubt of his exactneſs, though ſome of the appearances are ſuch as decompoſition and wearing might well enough be ſuppoſed to produce.

produce. The faireſt way is to take Mr Werner's obſervations juſt as they are given us, and to try whether they cannot be explained without the aſſiſtance of his theory. In effect, the wacken which he deſcribes, reſts, it would ſeem, on an unconſolidated bed of clay; and it may be ſuppoſed, that a part of this bed has been converted into wacken by the heat of the incumbent maſs, and has thus produced the apparent gradation from the one ſubſtance to the other. As the appearances of the rocks of Scheibenberg ſeem to be conſidered by Werner as furniſhing a very ſtrong, and even an unexpected confirmation of his ſyſtem, I cannot help thinking, that an explanation of them, on the principles of Dr Hutton, without any ſtraining or forcing of thoſe principles, contributes not a little toward extending the empire of the latter over all the phenomena of geology.

252. Another fact, which has been much inſiſted on of late, in proof of the aqueous formation of baſaltic rocks, is that ſhells are found in them. Of the reality of this fact, however, or at leaſt of the inſtances hitherto produced, great doubts I think may be reaſonably entertained. The ſpecimens of the ſuppoſed baſaltes, with ſhells included in them, that are chiefly relied on, are found at Portruſh in Ireland, a rocky promontory to the weſtward of the Giant's Cauſeway, and ſeparated from it by a

conſiderable

confiderable body of calcareous ftrata. Some of thefe fpecimens were brought to Edinburgh about a year ago, and were fuppofed, I believe, to contain an irrefragable proof of the Neptunian origin of the bafaltic promontory where they were found. I went to fee thefe fpecimens in company with Lord Webb Seymour and Sir James Hall; and, on examining them carefully, we were all of opinion, that the ftones which contained the fhells, or the impreffions of the fhells, were no part of the real bafaltes. They were all very compact, and had all more or lefs of a filiceous appearance, fuch as that of chert; they had nothing of a fparry or cryftallized ftructure; their fracture was conchoidal, and but flightly uneven. In two of them, one of which bore the impreffion of a *cornu ammonis*, the fchiftofe texture might be diftinctly perceived. A fpecimen which accompanied them, but in which there was no fhell, ferved very exactly to explain the relation between thefe ftones and the true bafaltes. Part of this fpecimen was a true bafalt, and the reft a fort of hornftone, exactly the fame with that in which the fhells were, and not unlike the jafper that is under the whinftone of Salifbury Crag, and in contact with it; fo that on the whole it was evident, that the rock containing the fhells is the fchiftus or ftratified ftone, which ferves as the bafe of the bafaltes, ſaltes, and which has acquired a high degree of induration, by the vicinity of the great ignited maſs of whinſtone.

This ſolution of the difficulty has ſince been confirmed by obſervations made on the ſpot by Dr Hope, who diſcovered two or three alternations of the baſaltic rock, with the beds of the ſchiſtus in which the ſhells are contained.

253. This alſo explains ſome obſervations of Spallanzani, made in the iſland of Cerigo, on the coaſt of Greece, the Cythæra of the ancients*. The baſe of that iſland is limeſtone; but it abounds alſo in unſtratified rocks, which the Italian naturaliſt ſuppoſes to be of volcanic origin; but which, if I miſtake not, we would regard as whinſtone, or perhaps porphyry; and they are ſaid to contain oyſter-ſhells and pectenites of a large ſize, perfectly mineralized. Theſe petrifactions, however, Spallanzani ſays, are not contained in the lava that has actually flowed, but in ſtones which have only endured a ſlighter action of fire. Without the commentary afforded by the Portruſh ſpecimens, it would be difficult to make out any thing very preciſe from this deſcription. By help of the information derived from thoſe ſpecimens, we may conclude, that the condition of the ſhells in

* Journal de Phyſique, tom. xlviii. (1798), p. 278.

in them, and in the rocks of Cerigo, is perfectly alike; and that, in both cafes, the fhells are involved in parts of the rock which are truly ftratified, but which have been, in fome degree, affimilated to the bafaltes by the heat which they have endured. Spallanzani would probably have ufed exactly the fame terms which he employs in fpeaking of Cerigo, if he had been required to defcribe the petrified fhells at Portrufh.

254. In the inftances juft mentioned, the petrified marine objects are not found in the real whinftone; but if they were found in it, when it borders on ftratified rocks containing fuch objects, the thing would not be at all furprifing, nor furnifh any argument againft the igneous confolidation of the ftone. If a torrent of melted matter was poured in among the ftrata, by a force which at the fame time broke up and difordered thofe ftrata, nothing could be more natural, than that this matter fhould contain fragments of them, and of the objects peculiar to them.

In one inftance, mentioned by Mr Strange, this feems actually to have taken place. In the Veronefe, a country remarkable for a mixture of limeftone ftrata, containing marine objects, with volcanic or bafaltine hills, he affures us, that he had feen a mafs of ftone, which had

 evidently

evidently concreted from fuſion, in which the marine foſſil bodies, originally, as he ſuppoſes, contained in the ſtrata, were perfectly diſtinguiſhable, though variouſly disfigured*. It may be, that in this, as in the foregoing examples, it was not real baſaltes, or real lava, which contained the ſhells, but the conterminal rock; but, ſuppoſing it to be as Mr Strange repreſents it, there appears to be no inconſiſtency between the phenomenon, and the igneous origin of the rock in which the ſhells were included. Here, however, it ſhould be remarked, that the preſence of great preſſure, to prevent the converſion of the ſhells into quicklime, ſeems abſolutely neceſſary; and that the phenomenon of theſe baſaltic petrifactions, requires the application of heat to have been deep under the ſurface of the earth.

255. The phenomena we have been conſidering, have been ſelected as the moſt unfavourable to the igneous origin of baſaltic rocks; and we have ſeen, that when duly examined, they are not at all inconſiſtent with it. We are now to take a view of ſome appearances, that ſeem quite irreconcilable with the aqueous formation of theſe rocks.

Where

* Phil. Tranſ. 1775, p. 25.

Where whinſtone rocks are found in maſſes, bounded by the ſtrata, and inſulated among them, they ſubject the Neptunian ſyſtem to great difficulties. For, ſuppoſing it true that this ſtone may be produced by the precipitation and cryſtallization of mineral ſubſtances diſſolved in water, yet it ſeems unaccountable, that this effect has been ſo local and limited in extent, as often to be confined to an irregular figure of a few acres, while, all round, the ſubſtances depoſited have had no tendency to cryſtallization, and have been formed into the common ſecondary ſtrata. The rock of Saliſbury *Craig*, for inſtance, is a maſs of whinſtone, having a perpendicular face eighty or ninety feet high toward the weſt, and extending from north to ſouth with a circular ſweep about 900 yards. The whole of this rock reſts on regular beds of ſecondary ſandſtone, not horizontal, but conſiderably depreſſed toward the north-eaſt: the rock is loftieſt in the middle, and decreaſes in thickneſs toward each end, terminating at its northern extremity in a kind of wedge. It is covered at top, toward that extremity, with regular beds of ſandſtone, perfectly ſimilar to thoſe on which it is incumbent; and it is not improbable, that this covering formerly extended over the whole.

 Now,

Now, what caufe can have determined the column of water, which refted on the bafe at prefent occupied by this rock, to depofite nothing but the materials of whinftone, while the water on the fouth, weft, and north, was depofiting the materials of arenaceous and marly ftrata? Wherefore, within this fmall fpace, was the precipitate every where *chemical*, to ufe the language of Werner, while clofe to it, on either fide, it was entirely *mechanical?* Why is there, in this cafe, no gradation? and why is a mere mathematical line the boundary between regions where fuch different laws have prevailed? Whence alfo, we may afk, has the bafaltic depofite been abruptly terminated toward the weft, fo as to produce the fteep face which has juft been mentioned? The operation of currents, or of any motion that can take place in a fluid, will furnifh no explanation whatever of thefe phenomena; yet they are phenomena far from being peculiar to a fingle hill; they are among the moft general and characteriftic appearances in the natural hiftory of whinftone mountains; and a geological theory which does not account for them, is hardly entitled to any confideration.

256. The bafaltic rock, juft defcribed, is alfo covered, at leaft partly, with ftrata perfectly fimilar

lar to thofe that lie under it. Now, it appears altogether unaccountable, that after the water had done depofiting the materials of the whin on the fpot in queftion, the former order was fo quickly refumed, and a depofition of fand, and of the other materials of the ftrata, took place juft as before. All this is quite unintelligible; and the principles of the Neptunian fyftem feem here to ftand as much in need of explanation, as any of the appearances which they are intended to account for.

257. The unequal thicknefs, and great irregularity in the furface of the whinftone mafs, here treated of, and of many rocks of the fame kind, is alfo a great objection to the notion of their aqueous formation. This feems to have been perceived by Werner, in the inftance of the rocks formerly mentioned; and he endeavours to explain it, by fuppofing, that much of thefe rocks has been deftroyed by wafte and decompofition, fo that an irregularity of their furface, and want of correfpondence has been given to them, which they did not originally poffefs. In the inftance of Salifbury *Craig*, however, we have a proof, that the great irregularity of furface, and the inequality of thicknefs, do not always arife from thefe caufes. The thinneft part of that rock, toward its northern extremity, is

 ftill

ftill covered by the ftrata in their natural place, and has been perfectly defended by them from every fort of wearing and decay. The cuneiform fhape, therefore, which this rock takes at its extremities, and the great difference of its thicknefs at them and in the middle, is a part of its original conftitution, and can be attributed to nothing cafual, or fubfequent to its confolidation.

The fame may be faid of many other bafaltic rocks, where an inequality of thicknefs, moft unlike to what belongs to aqueous depofites, is known to exift in beds of whinftone that are ftill deep under the furface. Thus the toadftone of Derbyfhire, even where it has a thick covering of ftrata over it, has been found, by the finking of perpendicular fhafts, to vary from the thicknefs of eighteen yards to more than fixty, within the horizontal diftance of lefs than a furlong. Nothing of this kind is ever found to take place in thofe beds of rock which are certainly known to originate from aqueous depofition, and no character can more ftrongly mark an effential difference of formation.

258. We have had frequent occafion to confider the characters of thofe maffes of whinftone which are fo often found interpofed between ftratified rocks. Thefe have been found in general very adverfe to the Neptunian fyftem ; and

two

two of them which yet remain to be mentioned, are even more ſo than any of the reſt.

Where a bed or tabular maſs of whinſtone is interpoſed between ſtrata, and wherever an opportunity offers of ſeeing its termination, if the ſtrata under it are not broken, it may be remarked, that they do not abut themſelves bluff and abrupt againſt the whin. On the contrary, if we mark the courſe of the ſtratum which covers the whinſtone, and of that which is the baſe of it, we ſhall find they converge toward one another, the interpoſed maſs growing thinner and thinner, like a wedge. When the latter terminates, the two former come in contact, and have no ſtratum interpoſed between them. Thus the roof and baſe of the whinſtone rock are contiguous beds, that appear as if they had been lifted up and bent, and ſeparated by an interpoſed maſs. Had the whole been an effect of ſimultaneous depoſition, the regular ſtrata muſt have been abruptly terminated by the whin, like two courſes of different ſorts of maſonry where they meet with one another.

259. From this wedge-form of the whinſtone maſſes, and in general from the irregularity of their ſurfaces, another concluſion follows, ſimilar to the preceding, and one which has been already mentioned. Where the ſurface of the interpo-

fed mafs is greatly inclined to the horizon, the ftrata which reft on this inclined plane, are nevertheless as exactly parallel to that plane, and to one another, as if they were really horizontal. It is certain, therefore, that they were not depofited on the fame inclined plane on which they now reft; for, if fo, they would have been ftill nearly horizontal, and by no means parallel to the inclined fide of the whinftone. This follows from the nature of aqueous depofition, as already explained.

We have a remarkable inftance of the phenomenon here referred to, in the rock of Salifbury *Craig*, of which mention has been fo often made, and in which almoft every circumftance is united, that can ferve to elucidate the natural hiftory of bafaltic rocks. The north end of that rock is in the figure of a wedge, with its inclined fide confiderably fteep, and covered by ftrata of grit, perfectly regular, and parallel to the furface on which they lie. The infpection of them will convince any one, that they were not depofited by the water, on a bottom fo highly inclined as that on which they now reft. They are of a ftructure very fchiftofe; their layers very thin; fo that any inaccuracy of their parallelifm would be readily obferved. The appearances of the horizontal depofition of thefe ftrata, are indeed fo clear, and fo impoffible

ble to be misunderstood, that the followers of the Huttonian system would not risk much, if they were to leave the whole theory of whinstone to the decision of this single fact, and should agree to abandon that theory altogether, if the Neptunists can shew any physical or statical principle, on which the deposition now described can possibly have been made ; or will point out the rule, by which nature has given a structure so nicely stratified to arenaceous beds deposited on a surface so highly inclined. If no such principle can be pointed out, though we cannot conclude that the Huttonian Theory is true, we certainly may conclude that the Neptunian is false.

260. Proofs of the igneous formation of whinstone, still more direct, are derived from the induration of the contiguous strata ; from their disturbance when intersected by veins of whinstone ; and from the charring of the coal which happens to be in contact with these veins. These are considered above at § 66, 67, &c. ; and it is particularly taken notice of at § 66, that pieces of sandstone are sometimes found as if floating in the whinstone, and, at the same time, greatly altered in their texture. One of the best and most unequivocal instances of this sort which I have seen, is to be found on the south side of *Arthur's Seat*, near Edinburgh. The rock which

which composes the upper part of the hill, on that side, is a whinstone breccia, such as we have many examples of, and, I believe, very much resembling what is called a *lava brecciata* by the volcanic geologists. The stony fragments included in this compound mass, are for the greater part rounded; and some of them are of whinstone, others of porphyry, strongly characterized by rectangular maculæ of feltspar, and many seem to be of sandstone, but so considerably altered, as to leave it at least disputable whether they really are so or not. In one part, however, where the face of the rock is nearly perpendicular, a narrow ridge is seen standing out from the rest, and of a different colour, being more entirely covered with moss than the rock round about it, and, as may be presumed from that circumstance, less liable to decomposition. On examination I found, that this ridge does not consist of whinstone, but of a very hard and highly consolidated sandstone. It appears to be the edge of a stratum, of the thickness of about nine or ten inches, and of the height of fifteen or sixteen feet. It is not perfectly straight, but slightly waved, its general direction being nearly vertical; and it is on both sides firmly embraced by the whinstone. When broken, it appears that this sandstone resembles in colour, and in every thing

but

but its greater conſolidation, and more vitreous ſtructure, the common grit found at the bottom of the hill, and over all the adjacent plain.

261. If all theſe circumſtances are put together, there appears but one concluſion that can be drawn from them. We have here the manifeſt marks of ſome power which could lift up this fragment of rock from its native place, diſtant at leaſt ſeveral hundred yards from its preſent ſituation, place it upright on its edge, encompaſs it with a ſolid rock, of a nature quite heterogeneous to itſelf, and beſtow on it, at the ſame time, a great addition of ſolidity and induration. If the maſs in which this ſtone is now imbedded, be ſuppoſed to have been once in fuſion, and forcibly thrown up from below, invading the ſtrata, and carrying the fragments along with it, the whole phenomena now deſcribed admit of an explanation, and all the circumſtances accord perfectly with one another; but, without this ſuppoſition, they are ſo many ſeparate prodigies, which have no connection with one another, nor with any thing that is known. It is indeed impoſſible, that the effects of motion and heat can be more clearly expreſſed than they are here, or the ſubject in which theſe powers reſided more diſtinctly pointed out.

262. The

262. The preceding facts being susceptible but of one interpretation, are on that account extremely valuable. The phenomena of Salisbury *Craig*, near the same place, are almost equally free from ambiguity. The basaltic rock which forms that precipice, rests on arenaceous or marly strata; and these, in their immediate contact with the former, afford an instance of what is mentioned § 67, namely, the conversion of the strata in such situations into a kind of petrosilex, or even jasper. The line which separates the one rock from the other, is, at the same time, so well defined, as, in the eyes even of the most determined Neptunist, to exclude all idea of insensible gradation.

263. The same rock affords some remarkable instances of the disturbance of the strata contiguous to the whinstone. The beds of the former are bent upwards in several places; and, at one in particular, form an arch, with its convexity downward, so as to make it evident, that the force which produced this bending was directed from below upwards.

264. It is, however, where whinstone takes the form of veins, intersecting the strata, that the induration of the latter is most conspicuous. The coast of Ayrshire, and the opposite coast of

Arran,

Arran, exhibit theſe veins in aſtoniſhing variety and abundance. The ſtrata are, in many inſtances, ſo *reticulated* by the veins, and interſected at ſuch ſmall diſtances, that it ſeems neceſſary to ſuppoſe, that the fiſſures in them were hardly ſooner made than filled up. This at leaſt is true, if the veins are to be accounted all of the ſame formation; and, in the greateſt number of inſtances by far, there is no mark of the one being poſterior to the other.

265. The induration of the ſides of theſe veins, in ſome caſes, has been ſuch, that the ſides have become more durable than the vein itſelf; ſo that the whinſtone has been worn away by the waſhing of the waves, and has left the ſides ſtanding up, with an empty ſpace, like a *ditch*, between them. One of theſe I remarked on the ſouth ſide of Brodick Bay, in Arran, which, where it met the face of an abrupt cliff, was not leſs than forty or fifty feet in depth.

266. I ſhall paſs over whatever argument might be drawn in favour of our ſyſtem, from the ſlender ramifications of the veins, and the varieties of their ſizes, from a few inches to many fathoms in diameter, and alſo from the connection which they often appear to have with the great tabular maſſes of baſaltes; and ſhall only add

add a few remarks on the charring of coal in the vicinity of veins or maſſes of whinſtone. The connection between the charring of coal and the preſence of whinſtone, was firſt obſerved by Dr Hutton; and, as far as opportunities of verifying the obſervation have yet occurred, appears to be a fact no leſs general than it is curious and intereſting. In the coal-mines of Scotland, it certainly holds remarkably, particularly in thoſe about Saltcoats in Ayrſhire, where a whinſtone dike is known to ſtretch acroſs the whole of the coal country, and to be every where accompanied with blind or uninflammable coal. At Newcaſtle, dikes of the ſame kind are met with, and one, in particular, in what is called the *Walker* Colliery, has proved the action of ſubterraneous fire, to the ſatisfaction of mineralogiſts nowiſe prejudiced in favour of the Huttonian ſyſtem.

The coal found under baſaltes, in the Iſland of Skye, has been already mentioned, § 139. To what was ſaid concerning the fibrous ſtructure of the parts of that foſſil in immediate contact with the whin, it may be added, that it is alſo charred in thoſe parts, ſo as to have hardly any flame when it is burnt, though further down it is of the nature of ordinary coal. Indeed, if there be any truth in Mr Kirwan's general remark,

mark, that it is common to find wood-coal under baſaltes, it muſt be underſtood to ariſe from this, that the coal in contact with the baſaltes is frequently charred, and its fibrous ſtructure, by that means, rendered more viſible.

267. It has been objected to the ſuppoſition of coal having its bituminous part driven off by the heat of the whinſtone, that this ought not, on Dr Hutton's principles, to happen in the mineral regions. But it may be replied, as has been done above, that the local application of heat might certainly produce this effect, and might drive off the volatile parts from a hotter to a colder part of the ſame ſtratum. The bitumen has not been ſo volatilized and expanded as entirely to eſcape from the mineral regions; but it has been expelled from ſome parts of a maſs, only to be condenſed and concentrated in others. This ſuppoſition coincides exactly with the appearances.

268. The native or foſſil-coke which accompanies whinſtone, has been diſtinguiſhed into two varieties. The firſt is the moſt common, in which, though the coal is perfectly charred, it is ſolid, and breaks with a ſmooth and ſhining ſurface. The ſecond is alſo perfect charcoal, but is very porous and ſpungy. This ſubſtance is much rarer than the other: Dr Hutton mentions an inſtance

inſtance of it at the mouth of the river Ayr, where there is a whinſtone dike *. I had the ſatisfaction of viſiting it along with him. It was in the bed of the river, below the high-water mark; the ſpecimens had the exact appearance of a *cinder*.

In the banks of the ſame river, ſome miles higher up, he found a piece of coal, belonging to a regular ſtratum, involved in whinſtone, and extremely incombuſtible. It conſumed very ſlowly in the fire, and deflagrated with nitre like plumbago. This he conſidered as the ſame foſſil which has been deſcribed under the name of *plombagine*. Near it, and connected with the ſame vein of whinſtone, was a real and undoubted plumbago.

From theſe circumſtances he alſo concluded, that plumbago is the extreme of a gradation, of which foſſil-coal is the beginning, and is nothing elſe than this laſt reduced to perfect charcoal. This agrees with the chemical analyſis, which ſhows plumbago to be compoſed of carbon, combined with iron.

In confirmation of this theory, he mentions a ſpecimen, in his poſſeſſion, of ſteatitical whinſtone, from Cumberland, containing nodules of a very perfect and beautiful plumbago; and he alſo takes notice of a mine of this laſt,

* Theory of the Earth, vol. i. p. 611.

laſt, in Ayrſhire, which, on the authority of Dr Kennedy, who has examined it with great care, I can ſtate as being contained, or enveloped in whinſtone; and I hope the public will ſoon be favoured with a particular deſcription of this very intereſting ſpot, by the ſame ingenious and accurate obſerver.

269. Thus the mineralogical and chemical diſcoveries agree in repreſenting coal, blind coal, plombagine, plumbago, as all modifications of the ſame ſubſtance, and as exhibiting the ſame principle, carbon, in a ſtate of greater or leſs combination. As the laſt and higheſt term of this ſeries ſhould be placed the *diamond;* but we are yet unacquainted with the matrix of this curious foſſil, and its geological relation to other minerals. When known, they will probably give to this ſubſtance the ſame place in the geological, as in the chemical arrangement: in the mean time, it is hardly neceſſary to remark, how well all the preceding facts agree with the hypotheſis of the igneous formation of whinſtone, and how anomalous and unconnected they appear, according to every other theory.

270. Notwithſtanding all this accumulated and unanſwerable evidence for the igneous formation of baſaltes, a great objection would ſtill remain to our theory, were it not for the very accurate and concluſive experiments concern-

ing the fuſion of this foſſil, referred to above, § 75. A ſtrong prejudice againſt the production of any thing like a real ſtone by means of fuſion, had ariſen, even among thoſe mineralogiſts, who were every day witneſſes of the ſtony appearance aſſumed by volcanic lava. They ſtill maintained, on the authority of their own imperfect experiments, that nothing but glaſs can ever be obtained by the melting of earths or of ſtones, in whatever manner they are combined.

An ingenious naturaliſt, after deſcribing a block of baſaltes, in which he diſcovered ſuch appearences, as inclined him to admit its igneous conſolidation, rejects that hypotheſis, merely from the imaginary inability of fire to give to any ſubſtance a ſtony character: "Quelque mélange," ſays he, "de terres que l'on ſuppoſe, quelque ſoit le degré de feu que l'on imagine, quelque ſoit le tems que l'on emploie, il eſt très certain que l'on n'obtiendra pas, par le ſeul fluide igné, ni baſalte, ni rien qui lui reſſemble *."

Sir James Hall's experiments have completely demonſtrated the contrary of what is here aſſerted; they have added much to the evidence of the Huttonian ſyſtem; and, independently of all

* Journal de Phyſ. tom. xlix. (1799) p. 36.

all theory, have narrowed the circle of prejudice and error.

Note xv. § 83.

On Granite.

1. *Granite Veins.*

271. It is ſaid above, § 77., that granite is found in unſtratified maſſes, and in veins. In the former of theſe conditions, it conſtitutes entire mountains, and forms the central ridge of many of the greateſt chains that traverſe the ſurface of the earth. It is the granite of this kind that has been moſt generally deſcribed by travellers and mineralogiſts. The veins have not been ſo much attended to, though they are of peculiar importance for aſcertaining the relation between granite and other foſſils.

272. Though Dr Hutton was the firſt geologiſt who explained the nature of granite veins, and who obſerved with attention the phenomena which accompany them, he is not the firſt who has mentioned them. M. Beſſon found veins of this kind in the Limoges, in an argillaceous ſchiſtus, and unconnected, as far as appeared, with any large maſs of granite *.

 Sauſſure

* Journal de Phyſ. tom. xxix. p. 89.

Sauſſure met with granite veins in the Valorſine, but did not ſee them diſtinctly. He aſcribed them to infiltration *. The date of this obſervation is in 1776: He afterwards diſcovered ſimilar appearances at Lyons †.

Werner alſo, in enumerating the ſubſtances of which veins are formed, reckons granite as one of them.

273. Veins of granite may be conſidered as of two kinds, according as they are connected, or not connected a parently with any large maſs of granite. It is probable, that theſe two kinds of veins only differ in appearance, and that both are connected with maſſes of the ſame rock, though that connection is viſible in ſome inſtances, and inviſible in others. The diſtinction, however, whatever it be with reſpect to the thing obſerved, is real with reſpect to the obſerver; and, as it is right, in a deſcription of facts, to avoid every thing hypothetical, I ſhall ſpeak of theſe veins ſeparately.

274 Veins of granite, having no communication, ſo far as can be diſcovered, with any maſs of the ſame rock, are found in the Weſtern Iſlands of Scotland, particularly in that of Coll, where

* Voyages aux Alpes, tom. i. § 598, 599.

† *Ibid.* § 601.

where they traverſe the beds of gneiſs and hornblend ſchiſtus, which compoſe the main body of the iſland. They are ſometimes ſeveral fathoms in thickneſs, obliquely interſecting the planes of the ſtrata juſt mentioned, which are nearly vertical. In theſe veins the feltſpar is predominant; it is very highly cryſtallized, and of a beautiful fleſh colour. Many ſmaller veins are alſo to be met with in the ſame place; but no large maſs of granite is found, either in this or the a acent iſland of Tiree.

275. The Portſoy granite, of which mention has been already made, § 80, alſo conſtitutes a vein or dike, traverſing a highly indurated micaceous ſchiſtus, about a mile to the eaſtward of the little town of Portſoy, and not viſibly connected with any large maſs of the ſame kind. More dikes than one of this granite have been obſerved ear the ſame ſpot.

A ſimilar granite is likewiſe found inland, in the neighbourhood of Huntly, about eighteen miles ſouth of Portſoy; but whether in the ſhape of a vein or a maſs, I have not been able to learn.

276. Veins of granite are alſo frequent in Cornwall, where they are known by the name of *lodes*, the ſame name which is applied in that country to metallic veins. The granite veins fre-

quently interſect the metallic, and are remarkable for producing ſhifts in them, or for throwing them out of their natural direction. The mineral veins, particularly thoſe that yield copper and tin, run nearly from eaſt to weſt, having the ſame direction with the beds of the rock itſelf, which is a very hard ſchiſtus. The granite lodes, as alſo thoſe of porphyry, called *elvan* in Cornwall, are at right angles nearly to the former; and it is remarked, that they generally heave the mineral veins, but that the mineral veins ſeldom or never heave the croſs-veins. In this country, therefore, the veins of granite and porphyry are poſterior in formation to the metallic veins. Theſe veins of granite may perhaps be connected with the great granitic maſs that runs longitudinally through Cornwall, from Dartmoor to the Land's End. This much is certain, that their directions in general are ſuch, that, if produced, they would interſect that maſs, nearly at right angles.

277. The granite veins in Glentilt, where Dr Hutton made his firſt obſervations on this ſubject, are not, I believe, viſibly connected with any large maſs of the ſame rock *. The bed of the river Tilt, in the diſtance of little more than a mile, is

* Tranſ. Royal Society Edin. vol. iii. p. 77, &c.

is intersected by no less than six very powerful veins of granite, all of them accompanied with such marks of disorder and confusion in the strata, as indicate very strongly the violence with which the granite was here introduced into its place. These veins very probably belong to the great mass of granite which is known to form the central ridge of the Grampians further to the north; but they are several miles distant from it, and the connection is perhaps invisible in the present state of the earth's surface.

278. The second kind of granite vein, is one which proceeds visibly from a mass of that rock, and penetrates into the contiguous strata. The importance of this class of veins, for ascertaining the relation between granite and other mineral bodies, has been pointed out, § 82.; and by means of them it has been shewn, that the granite, though inferior in position, is of more recent formation than the schistus incumbent on it; and that the latter, instead of having been quietly deposited on the former, has been, long after its deposition and consolidation, heaved up from its horizontal position, by the liquid body of granite forcibly impelled against it from below.

It has been alleged, in order to take off the force of the argument derived from granite veins, that theſe veins are formed by infiltration, though, to give any probability to this ſuppoſition, it would be neceſſary to ſhew, that water is able to diſſolve the ingredients of granite; and even if this could be done, the direction which the veins have, in many inſtances, riſing up from the granite, is a proof, as remarked § 82., that they cannot be the effect of infiltration.

Another objection has been thrown out, namely, that the veins here referred to are not of true granite, according to the definition which mineralogiſts have given of that ſubſtance. The force of a fact, however, is not to be leſſened by a change of names, or the uſe of arbitrary definitions. The general fact is, that the granitic maſs, and the vein proceeding from it, conſtitute one continuous, and uninterrupted body, without any line of ſeparation between them. The geological argument turns on this circumſtance alone; and it is no matter whether the rock be a ſyenite, a granitelle, or a real granite. The phenomenon ſpeaks the ſame language, and leads to the ſame concluſion, whatever be the technical terms the mineralogiſt employs in deſcribing it.

279. It muſt, however, be admitted, that a difference in character is often to be obſerved between the granite maſs and the veins proceeding

ing from it; ſometimes the ſubſtances in the latter are more highly cryſtallized than in the former; ſometimes, but more rarely, they are leſs cryſtallized, and, in ſome inſtances, an ingredient that enters into the maſs ſeems entirely wanting in the vein. Theſe varieties, for what we yet know, are not ſubject to any general rule; but they have been held out as a proof, that the maſſes and the veins are not of the ſame formation. It may be anſwered, that a perfect ſimilarity between ſubſtances that, on every hypotheſis, muſt have cryſtallized in very different circumſtances, is not always to be looked for; but the moſt direct anſwer is, that this perfect ſimilarity does ſometimes occur, inſomuch that, in certain inſtances, no difference whatſoever can be diſcovered between the maſs and the vein, but they conſiſt of the ſame ingredients, and have the ſame degree of cryſtallization. Some inſtances of this are juſt about to be remarked.

280. A ſtrong objection to the ſuppoſed origin of granitic veins from infiltration, and indeed to their formation in any way but by igneous fuſion, ariſes from the number of fragments of ſchiſtus, often contained, and completely inſulated in thoſe veins. How theſe fragments were introduced into the fiſſures of the ſchiſtus, and ſuſtained till they were ſurrounded by

by the matter depoſited by water, is very hard to be conceived; but if they were carried in by the melted granite, nothing is more eaſily underſtood.

The following are ſome of the places where the phenomena of granite veins may be diſtinctly ſeen.

281. The iſland of Arran, remarkable for collecting into a very ſmall compaſs a great number of the moſt intereſting facts of geology, exhibits many inſtances of the penetration of ſchiſtus by veins of granite. A group of granite mountains occupies the northern extremity of the iſland, the higheſt of which, Goatfield, riſes nearly to the higheſt of 3000 feet, and on the ſouth ſide is covered with ſchiſtus to the height of 1100. From thence, the line of junction, or that at which the granite emerges from under the ſchiſtus, winds, ſo far as I was able to obſerve, round the whole group of monntains, with many wavings and irregularities, riſing ſometimes to a greater, and deſcending ſometimes to a much lower level, than that juſt mentioned. Along this line, particularly on the ſouth, wherever the rock is laid bare, and cut into by the torrents, innumerable veins of granite are to be ſeen entering into the ſchiſtus, growing narrower as they advance into it; and being directed, in very many caſes, from below

below upwards, they are precifely of the kind which the infiltration of water could not produce, even were that fluid capable of diffolving the fubftances which the vein confifts of. From this fouth face of the mountain, and from the bed of a torrent that interfects it very deeply, Dr Hutton brought a block of fchiftus, of feveral hundredweight, curioufly penetrated by granite veins, including in them many infulated fragments of the fchiftus.

From this point, the common fection of the granite and fchiftus defcends towards the weft fide of the mountain, and is vifible at the bottom of a deep glen, (Glen-Rofa), which detaches Goatfield from the hills farther to the weft. The junction is laid bare at feveral places in the bed of the river which runs in the bottom of this glen; and in all of them exhibits, in a greater or lefs degree, the appearances of difturbance and violence which have accompanied the injection of the granite veins. Many circumftances render this fpot interefting to a geologift, and, among others, an interfection of the granite, a little above its junction with the fchiftus, by a dike or vein of very compact whinftone.

The fame line of junction is found on the oppofite, or north-eaft, fide of the mountain, where

where it is interſected by another little river, the Sannax, which on this ſide determines the baſe of the mountain. This junction is no leſs remarkable than the other two.

The iſland of Arran contains, I have no doubt, many other ſpots where theſe phenomena are to be ſeen; but I have had no opportunity of obſerving them, nor do I find that Dr Hutton met with any others in his viſit to this iſland.

282. Another ſeries of granite veins is found in Galloway, which was firſt diſcovered by Dr Hutton and his friend Mr Clerk, and afterwards more fully explored by Sir James Hall and Mr Douglas, the preſent Earl of Selkirk. The two laſt traced the line of ſeparation between a maſs of granite and the ſchiſtus incumbent upon it, all round a tract of country, about eleven miles by ſeven, extending from the banks of Loch Ken weſtward; and in all this tract they found, "that wherever the junction of the granite with the ſchiſtus was viſible, veins of the former, from fifty yards, to the tenth of an inch in width, were to be ſeen running into the latter, and pervading it in all directions, ſo as to put it beyond all doubt, that the granite of theſe veins, and conſequently of the

the great body itſelf, which was obſerved to form with the veins one uninterrupted maſs, muſt have flowed in a ſoft or liquid ſtate into its preſent poſition *." I have only further to add, that ſome of theſe veins are remarkable for containing granite, not ſenſibly different, in any reſpect, from the maſs from which they proceed.

283. In Invernefsſhire, between Bernera and Fort Auguſtus, the ſame phenomena occur on the north ſide of Loch Chloney, where ſome granite mountains riſe from under the ſchiſtus. In travelling near this place, Lord Webb Seymour and myſelf were advertiſed of our approach to a junction of granite and ſchiſtus, by finding among the looſe ſtones on the road many pieces of ſchiſtus, interſected with veins of feltſpar and granite. We walked along this junction for more than a mile; and toward the eaſt end, where the road leaves it, we ſaw, in the bed of a ſtream that runs into Loch Chloney, many beautiful ſpecimens of granitic veins pervading the ſchiſtus, and branching out into very minute ramifications.

284. The laſt inſtance I have to mention from my own obſervation, is at St Michael's

* Tranſ. Royal Society Edin. vol. iii. p. 8.

chael's Mount in Cornwall. That mount is entirely of granite, thruſt up from under a very hard micaceous ſchiſtus, which ſurrounds it on all ſides. At the baſe of it, on the weſt ſide, a great number of veins run off from the granite, and ſpread themſelves like ſo many roots fixed in the ſchiſtus: they are ſeen at low water. In the ſmaller veins, the granite is of very minute, though diſtinct parts; in the larger, it is more highly cryſtallized, and is undiſtinguiſhable from the maſs of the hill.

Beſides the above, Cornwall probably affords many other inſtances of the ſame kind, which I have not had an opportunity to examine. Such inſtances may in particular be looked for at the Land's End, where a promontory, conſiſting of a central part of granite, and covered by micaceous ſchiſtus on both ſides of it, is cut tranſverſely by the ſea-coaſt, and the contact of the granite and ſchiſtus of courſe twice expoſed to view.

285. Scotland alſo affords other examples of granite veins, and ſome of them have been actually deſcribed. Mr Jamieſon has taken notice of ſome which he ſaw in the bottom of the river Spey, at Glen Drummond, in Badenach, and has repreſented them in an engraving.

graving*. They traverſe the ſtrata in various directions, and incloſe pieces of the micaceous ſchiſtus; and, from the great number of looſe blocks which he found, exhibiting portions of ſuch veins, it is probable, that they are very numerous in this quarter. The ſame mineralogiſt mentions ſome inſtances of ſimilar veins in the Shetland Iſles †.

In Roſs-ſhire, Sir George Mackenzie has obſerved a great variety of granite veins, ſome of them of large ſize. One of them, in particular, not far from Coul, when firſt diſcovered, was ſuppoſed to be a ſingle maſs, riſing from under the ſchiſtus; but, on a more careful examination, has been found to be a part of a great ſyſtem of veins, which interſects the micaceous ſchiſtus of this tract in various directions.

286. The granite veins are not the only proof that this ſtone is more recent than ſome other productions of the mineral kingdom. Specimens of granite are often found, containing round nodules of other ſtones, as, for example, of gneiſs or micaceous ſchiſtus. Such is the ſpecimen of granite containing gneiſs, which Werner himſelf is ſaid to be in poſſeſſion of, and to

* Mineralogy of the Scottiſh Iſles, vol. ii. p. 173.

† *Ibid.* p. 216.

to confider as a proof, that the fchiftus is of greater antiquity than the granite. Such alfo feemed to me fome pieces of granite, which I met with in Cornwall, near the Land's End; and others which I faw in Ayrfhire, in loofe blocks, on the fea-coaft between Ayr and Girvan. It is impoffible to deny that the containing ftone is more modern than the contained. The Neptunifts indeed admit this to be true, but allege, that all granite is not of the fame formation; and that, though fome granite is recent, the greater part boafts of the higheft antiquity which belongs to any thing in the foffil kingdom. This diftinction, however, is purely hypothetical; it is a fiction contrived on purpofe to reconcile the fact here mentioned with the general fyftem of aqueous depofition, and has no fupport from any other phenomenon.

2. *Granite of Portfoy.*

287. The granite of Portfoy is one of the moft fingular varieties of this ftone, and is remarkable for this circumftance, that the feltfpar is the fubftance which has affumed the figure of its proper cryftal, and has given its form

to

to the quartz, ſo that the latter is impreſſed both with the acute and obtuſe angles belonging to the rhombic figure of the former. The angular pieces of quartz thus moulded on the feltſpar, and ranged by means of it in rows, give to this ſtone the appearance of rude alphabetical writing.

Now, Dr Hutton argued, that ſubſtances precipitated from a ſolution, and cryſtallizing at liberty, cannot be ſuppoſed to impreſs one another in the manner here exemplified; and that they could do ſo only when the whole maſs acquired ſolidity at the ſame time, or at the ſame time nearly *. Such ſimultaneous conſolidation can be produced in no way that we know of, but by the cooling of a maſs that has been in fuſion.

288. A granite, brought from Daouria by M. Patrin, and deſcribed by him in the Journal de Phyſique for 1791, p. 295, under the name of *pierre graphique*, ſeemed to Dr Hutton to have ſo great a reſemblance to the granite of Portſoy, that he ventured to conſider them both as the ſame ſtone, and as both containing quartz moulded on feltſpar †. It ſhould ſeem, however,

X

* Theory of the Earth, vol. i. p. 104.

† Tranſ. Royal Society Edin. vol. iii. p. 83.

ever, from further explanations, which M. Patrin has ſince given, that Dr Hutton was miſtaken in his conjecture, and that, in the *pierre graphique* of the former mineralogiſt, the quartz gives its form to the feltſpar, preſerving in its cryſtals their natural angle of 120 degrees*. It is impoſſible, I think, to doubt of the accuracy of this ſtatement; and the graphical ſtone of Portſoy muſt therefore be admitted to differ materially from that of Daouria. They are not, however, without ſome conſiderable affinity, beſides that of their outward appearance; for, though the quartz in the former is generally moulded on the feltſpar, the feltſpar is alſo occaſionally impreſſed by the quartz, and ſometimes even included in it. They may be conſidered as varieties of the ſame ſpecies of granite; and the *pierre graphique* of Corſica is probably a third variety, different from them both.

289. It would ſeem, however, that all theſe ſtones lead exactly to the ſame concluſion. M. Patrin deſcribes his ſpecimen as containing quartz cryſtals, that are for the moſt part only *caſes*, having their interior filled with feltſpar. " Le feltſpath

* Journal Britannique (of Geneva), 1798, vol. viii. Sciences et Arts, p. 78.

ſpath en maſſe contient des cryſteaux quartzeux, qui n'ont le plus ſouvent que le carcaſſe, et dont l'interieur eſt rempli de feltſpath; ſouvent il manque à ces carcaſſes quelques unes de leurs faces, et ſouvent la ſection de cette pierre dans un ſens tranſverſal aux cryſtaux, preſente une ſuite de figures qui ſont des portions d'hexagones, et qui ne reſemblent pas mal à des caractères Hebraiques *."

Theſe imperfect hexagonal caſes of quartz, filled with feltſpar, certainly indicate the cryſtallization of ſubſtances, which all aſſumed their ſolidity at the ſame time, and, in doing ſo, conſtrained the figures of one another. To uſe the words of Dr Hutton, " whether cryſtallizing quartz incloſe a body of feltſpar, or concreting feltſpar determine the ſhape of fluid quartz, particularly if we have, as is here the caſe, two ſolid bodies including and included, it amounts to a demonſtration, that thoſe bodies have concreted from a fluid ſtate of fuſion, and have not cryſtallized, in the manner of ſalts, from a ſolution †."

290. The quartz in granite ſo generally receives the impreſſions of all the other ſubſtances,

 particularly

* Journal Britannique, *ibid.*

† Tranſ. Royal Society Edin. *ubi ſupra*, p. 84.

particulary of the feltſpar and ſchorl, and appears to be ſo paſſive a body, that it has been doubted by ſome mineralogiſts, whether in this ſtone it ever aſſumes its own figure, except where cavities afford room for its cryſtallization. But it is certain that, beſide the Daourian granite juſt mentioned, there are others, in which the quartz is completely cryſtallized. Of this ſort are ſome ſpecimens, found in a granite vein on the weſt ſide of the hill of St Agnes, in Cornwall. The vein traverſes the primitive ſchiſtus, of which that hill conſiſts, from ſouth to north nearly: the ſtone is much decompoſed, and the feltſpar in general is almoſt reduced to the ſtate of clay. In this decompoſed maſs, quartz cryſtals are found, having the ſhape of double hexagonal pyramids, perfectly regular and complete. The ſide of the hexagon, which is the baſe of the two oppoſite pyramids, varies from half a tenth to a tenth of an inch in length, and is the ſame with the altitude of each of the pyramids. In ſome few ſpecimens, the two pyramids do not reſt on the ſame baſe, but are ſeparated by a very ſhort, though regular, hexagonal priſm. The ſurfaces of theſe cryſtals are rough, and ſomewhat opaque, with ſlender ſpiculæ of ſhorl frequently traverſing them. This roughneſs is occaſioned by ſlight furrows

furrows on the ſurface of the cryſtal, very regularly diſpoſed, and parallel to one another, being without doubt impreſſions from the thin plates of the feltſpar, which ſurrounded the cryſtal, and ſlightly indented it. They very much reſemble ſome impreſſions, remarked by Dr Hutton in the granite of Portſoy, and aſcribed by him alſo to a ſimilar cauſe. He has repreſented theſe in his Theory of the Earth, vol. i. plate 2. fig. 4. The action and reaction of two cryſtallizing bodies, hardly admits of a ſtronger and more unequivocal expreſſion, than in theſe two inſtances.

Where the granite was little decompoſed, the quartz was not eaſily diſengaged from the maſs it was imbedded in, and often broke in pieces before it could be extricated. The cryſtallization of the quartz, therefore, would not have been diſcovered, but for the decompoſition of the feltſpar; and it is probable, that ſimilar cryſtallizations exiſt in many granites where they are not perceived.

291. Some mineralogiſts are inclined to think, that the regular cryſtallization of quartz is to be found only in what they call ſecondary granites, or in thoſe that are of a formation ſubſequent to the great maſſes which conſtitute the granite mountains. It is indeed true, that in the in-

ſtances given here, both from Cornwall and Daouria, the granites containing quartz-cryſtals are from veins that interſect the primary ſchiſtus, and are therefore, on every hypotheſis, of a formation ſubſequent to that ſchiſtus. But it does not follow from thence, that they are leſs ancient than the great maſſes of unſtratified granite; with theſe laſt they are moſt probably coëval, nor can there be any reaſon for thinking the cryſtallization of quartz a mark of more recent formation than that of feltſpar.

3. *Stratification of Granite.*

292. What are the various modes in which granite exiſts, is a queſtion not abſolutely decided among mineralogiſts. 1. That it exiſts as a ſchiſtoſe ſtone of a fiſſile texture, in gneiſs and *veined granite*, is on all hands admitted, though in this ſtate the name of granite is generally withheld from it. 2. That it exiſts often without any indication of a fiſſile texture, and altogether unſtratified, is likewiſe acknowledged. 3. That it is found in veins, interſecting the ſtrata, has been ſhown above. The only mode of its exiſtence ſubject to diſpute, is that in which it is ſaid to be ſtratified in its outward

ward configuration, but not ſchiſtoſe in its texture. On this point mineralogiſts do not perfectly agree: Dr Hutton did not think that this was a ſtate in which granite ever appears. When not ſchiſtoſe in its ſtructure, he ſuppoſed it to be unſtratified altogether; and he conſidered it as a body which, like whinſtone, was originally in a ſtate of igneous fuſion, and, in that condition, injected among the ſtrata. The ſchool of Werner, on the other hand, maintain, that granite, if not always, is generally ſtratified, and diſpoſed in beds, ſometimes horizontal, though more frequently vertical, or highly inclined.

In forming an opinion where there are great authorities on oppoſite ſides, a man muſt truſt chiefly to his own obſervations, and ought to eſteem himſelf fortunate if theſe lead to any certain concluſion. Mine incline me to differ from Dr Hutton, on the one hand, and from the Neptuniſts on the other, as they convince me, that granite does form ſtrata where it has no character of gneiſs; and, at the ſame time, induce me to ſuſpect, that the ſtratification aſcribed by the Neptuniſts to the granite mountains, is, in many inſtances, either an illuſion, or at leaſt ſomething very different from what, in other ſtones, is accounted ſtratification.

293. The firſt example I ever ſaw of granite that was ſtratified, and yet had no character of gneiſs, was at Chorley Foreſt, in Leiceſterſhire. The greater part of that foreſt has for its baſe a hornſtone ſchiſtus, primary and vertical; and, on its eaſtern border, particularly near Mount Sorrel, are beds of granite, holding the ſame direction with thoſe of the ſchiſtus. The ſtone is a real granite; it has nothing in its internal ſtructure of a ſchiſtoſe or fiſſile appearance; and its beds, which it is material to remark, are no thicker than thoſe of the hornſtone ſtrata in the neighbourhood. This granite is remarkable too, for being cloſe to the ſecondary ſandſtone ſtrata; I did not ſee their contact, but traced them within a ſmall diſtance of one another; ſo that I think it is not likely that any body of rock intervenes. At the ſame time that I ſtate my belief of this rock of granite being in regular ſtrata, I muſt acknowledge, that a very intelligent mineralogiſt, who viewed theſe rocks at the ſame time, and whoſe eye was well practiſed in geological obſervation, remained in doubt concerning them.

294. Another inſtance of a real granite, diſpoſed in regular beds, but without any character of gneiſs, is one which I ſaw in Berwickſhire, in Lammermuir, near the village of Prieſtlaw. The little river of Faſſnet cuts the beds acroſs, and

and renders it eaſy to obſerve their ſtructure. The beds are not very thick; they run from about S. S. W. to N. N. E. like the ſchiſtus on either ſide of them. I was in company with Sir James Hall when I ſaw theſe rocks; we examined them with a good deal of attention, and traced them for more than a mile in the bed of the river; and, if I miſtake not, our opinions concerning them were preciſely the ſame.

295. What exiſts in two inſtances may exiſt in many, and, after theſe obſervations, I ſhould be guilty of great inconſiſtency, in refuſing to aſſent to the accounts of Pallas, De Luc, Sauſſure, and many other mineralogiſts, who ſo often repreſent granite as formed into ſtrata. In ſome caſes, however, it is certain, that the ſtratification they deſcribe is extremely unlike that in the two inſtances juſt mentioned, and indeed very unlike any thing that is elſewhere known by the name of ſtratification. For example, the ſtratification muſt be very ambiguous, and very obſcurely marked, that was not diſcovered till after a ſeries of obſervations, continued for more than twenty years, by a very ſkilful and diſtinguiſhing mineralogiſt. Yet ſuch undoubtedly is the ſtratification of Mont Blanc, and of the granite mountains in its neighbourhood, as it eſcaped the eyes of Sauſſure, in the repeated viſits which he made to them, during a period of

no leſs extent than has juſt been mentioned. It was not till near the concluſion of thoſe labours, to which the geologiſts of every age will conſider themſelves as highly indebted, that, having reached the ſummit of Mont Blanc, he perceived, or thought that he perceived, the ſtratification of the granite mountains. The *Aiguilles* or Needles which border the valley of Chamouni, and even Mont Blanc itſelf, appeared to be formed of vaſt tabular maſſes of granite, in poſition nearly vertical, and ſo exactly parallel, that he did not heſitate to call them by the name of ſtrata. Till this moment, theſe ſame mountains, viewed from a lower point, had been regarded by him as compoſed of great plates of rock, nearly vertical indeed, but applied, as it were, round an axis, and reſembling the leaves of an artichoke*; and the fiſſures by which they are ſeparated from one another, had been conſidered as effects of waſte and degradation. " But now," (ſays he, ſpeaking of the view from the top of Mont Blanc), " I was fully convinced, that theſe mountains are entirely compoſed of vaſt plates of granite, perpendicular to the horizon, and directed from N. E. to S. W. Three of theſe plates, ſeparated from each other, formed the top

* Voyages aux Alpes, tom. ii. § 910, &c.

top of the *Aiguille du Midi*, and other ſimilar plates, decreaſing gradually in height, compoſe its declivity to the ſouth *."

296. Sauſſure was ſo ſtrongly impreſſed with the appearances of what he accounted regular ſtratification, ſuch as water only can produce, and ſuch as muſt have been in the beginning horizontal, that, placed as he now was, on one of the higheſt points of the earth's ſurface, he formed the bold conception, that the ſummit on which he was ſtanding had been once buried under the ſurface, to the depth at leaſt of half the diameter of the mountain, and horizontally diſtant from its preſent place by a line not leſs than the whole height of the mountain; the granite beds which compoſe that mountain, having been raiſed by ſome enormous power from their horizontal poſition, and turned as on an axis, till they were brought into the vertical plane. In this notion, which ſuits ſo well with the nature of mountains really compoſed of vertical ſtrata, and which does credit to the extent of Sauſſure's views, it is wonderful that he did not ſee the overthrow of the geological ſyſtem he had adopted, which is provided with no means whatſoever of explaining theſe great effects.

Such,

* Voyages aux Alpes, tom. iv. § 1996.

Such, then, were the ideas ſuggeſted to Sauſſure, by viewing the mountains of the Alps from the higheſt of their ſummits. His great experience, his accurate knowledge of the objects before him, and the power he had acquired of diſſipating thoſe illuſions, to which, in viewing mountainous tracts, the eye is peculiarly ſubject, all conſpire to give great weight to his opinion. Yet, as this opinion is oppoſed by that which he himſelf had ſo long entertained, before it can be received with perfect confidence, it will require to be verified by new obſervations. It ſeems certain, that the beds of rock here deſcribed, differ from all ordinary ſtrata, both horizontal and vertical, in the circumſtance of their vaſt thickneſs, three of them being ſo large as to form the main body of a mountain. Their paralleliſm cannot eaſily be aſcertained; and they have at beſt but a very ſlight reſemblance to ſuch beds as water is known to produce.

297. Their paralleliſm is difficult to be aſcertained; for, on account of the magnitude and inacceſſibility of the objects, it is impoſſible to place the eye in any ſituation, where it ſhall not be much nearer to one part of the planes whereof the paralleliſm is to be eſtimated, than to another. Indeed, one can perceive a cauſe which

which may have rendered the parallelifm of the plates of granite which compofe the *aiguilles*, more accurate in appearance than in reality, when viewed from a point fo elevated as the fummit of Mont Blanc. For, even on the fuppofition that the comparifon of thofe plates to leaves of artichokes was juft, and that the planes of their feparation converged toward one another, in afcending to the top, when they were viewed from a point more elevated than that top, this convergency would be diminifhed, and, by the force of the perfpective, might even be converted into parallelifm. We cannot at prefent afcertain what effect this caufe of deception may have actually produced.

298. The obfervations of Sauffure concerning the ftratification of granite, are not, however, in all inftances, liable to thefe objections; and it feems to be on much lefs exceptionable grounds that he pronounces the granite of St Gothard to be ftratified. The gneifs and micaceous fchiftus which conftitute the lower part of that mountain, are fucceeded by a granite without any fchiftofe appearance, but divided into large plates, exactly parallel to the beds of the former gneifs. Thefe he regards as real ftrata. On ftudying them in detail, he fays, confiderable irregularities were to be obferved, but not greater than in the cafe

of

of limeſtone or micaceous ſchiſtus *. It may be inferred from this, that theſe plates of granite are not ſo thick but that they admit of comparifon with beds that are known with certainty to be of aqueous formation, and I am therefore diſpoſed to believe, that the granite of St Gothard, in this part at leaſt, is ſtratified. The tranſition from gneiſs to granite *en maſs*, is not uncommon, as Sauſſure has obſerved in other inſtances, and as we are juſt about to conſider more particularly.

299. In the mountains of our own country, ſome difficulties concerning the ſtratification of granite have alſo occurred. In Arran, for inſtance, the mountain of Goatfield, which I have mentioned above as affording an inſtance of granite ſending out many veins into the ſchiſtus, and rivetted, as it were, by means of them to the ſuperincumbent rock, when I viſited it, with a view of verifying on the ſpot the intereſting obſervations which Dr Hutton had there made, appeared to me to be without any veſtige of ſtratification in its granitic part, as did alſo the whole group of mountains to which it belongs. It was, therefore, not without a good deal of ſurpriſe, that I lately read, in an account of that iſland, by a very accurate and ingenious mineralogiſt,

* Voyages aux Alpes, tom. iv. § 1830.

ralogiſt, that Goatfield conſiſts of ſtratified granite *. The impreſſion which the appearance of that mountain made on my mind, is juſt the reverſe; and though I ſaw large tabular maſſes, ſometimes nearly vertical, ſeparated by fiſſures, they appeared to be much too irregular, too little extended in length and height, and vaſtly too much in thickneſs, to be reckoned the effects of ſtratification. For all this, I would by no means be underſtood to ſet my obſervations in oppoſition to thoſe of Mr Jamieſon. In my viſit to Arran, I did not direct my inquiries much toward this point; the general appearance of the rocks did not ſuggeſt the neceſſity of doing ſo, and I was not perfectly aware how much the ſtratification of granite had been inſiſted on by ſome mineralogiſts; ſo that I applied myſelf entirely to ſtudy ſome other of the intereſting phenomena which this little iſland offers in ſo great abundance. I therefore carry my confidence in the appearances which ſeemed to indicate a want of ſtratification in the granite of Arran no further than to remain ſceptical both as to Mr Jamieſon's concluſions and my own, till an opportunity

* Mineralogy of the Scottiſh Iſles, vol. i. p. 35, 36.

tunity ſhall occur of verifying the one or the other by actual obſervation.

300. The ſtratification of granite, though it made no part of Dr Hutton's ſyſtem, does by no means embarraſs his theory with any new difficulty. Rocks, of which the parts are highly cryſtallized, are already admitted as belonging to the ſtrata, and are exemplified in marble, gneiſs, and veined granite. In the two laſt, we have not only ſtratification, but a ſchiſtoſe, united with a cryſtallized ſtructure, and the effects of depoſition by water, and of fluidity by fire, are certainly nowhere more ſingularly combined. The ſtratification of theſe ſubſtances is therefore more extraordinary than even that of the moſt highly cryſtallized granite. Neither the one nor the other can be explained but by ſuppoſing, that while ſuch a degree of fluidity was produced by heat, as enabled the body when it cooled to cryſtallize, the whole maſs was kept in its place by great preſſure acting on all ſides, ſo that the ſhape was preſerved as originally given to it by the ſea. As we cannot, however, ſuppoſe, that the intenſity of the heat, or the fuſibility of the ſubſtance through all the parts of a ſtratum, were preciſely the ſame, we may expect to find in the ſame ſtratum, or in the ſame body of ſtrata, that in ſome parts the marks of ſtratification are

completely

completely obliterated, while in others they remain entire. It is thus that *veined granite*, or what I think ſhould be called granitic ſchiſtus, often graduates into granite *in maſs*, that is, granite without any ſchiſtoſe or fiſſile texture. Sauſſure ſays, that to be veined or not veined, is an affection of granite, that ſeems, in many caſes, accidental *; as, in the midſt of rocks of that ſubſtance, moſt clearly fiſſile, large portions appear without any veſtige of ſtratification. Of this phenomenon, which is frequent in the Alps, inſtances are alſo to be met with in the granite rocks of Scotland, and the adjacent iſles; and I know that Dr Hope, in a mineralogical excurſion which he lately made among the Hebrides, obſerved many intereſting and curious examples of it. Indeed, when rocks were ſo much fuſed as to cryſtallize, and ſo compreſſed, at the ſame time, as to remain ſtratified, they were evidently on the verge of change; two oppoſite forces were very nearly balanced, and each carried as far as it could go without entirely overcoming the other; ſo that a ſmall alteration in the conditions may have made a great alteration in the effects. Hence a ſudden tranſition from a ſtratified to

Y an

* Voyages aux Alpes, tom. iv. § 2143.

an unſtratified texture, which is only found in rocks highly cryſtallized, and ſuch as have endured the moſt violent action of the mineralizing powers.

301. Now, though the ſtratification of granite, or the mixture of the ſtratified with the unſtratified rocks of that genus, is not only reconcileable with the principles of the Huttonian geology, but might even have been deduced as a corollary from thoſe principles, before it was actually obſerved, it may be conſidered as inconſiſtent with the theory of granitic veins that has juſt been given. A ſtratum, though ſoft or fluid, could not invade the ſurrounding ſtrata with violence, nor ſend out veins to penetrate into them. It might, if ſtrongly compreſſed by another ſtratum leſs fluid than itſelf, fill up any fiſſures or cracks that were in that other, but this would hardly produce ſuch large veins, and of ſuch conſiderable length, as often penetrate from the granite into the ſchiſtus, nor could it give riſe to any appearance of diſturbance. If, therefore, veins were found proceeding from ſuch ſtratified granite as that of Chorley Foreſt or Lammermuir, I ſhould think, that the explanation of them was ſtill a *deſideratum* in geology. The Neptunian theory of infiltration would indeed be as applicable to them

them as to any other veins; for it is but little affected by the condition of the phenomena to be explained. Indeed, it is very difficult to set any limits to the explanations which this theory affords; and it would certainly puzzle a Neptunist, to assign any good reason why infiltration has not produced veins of one schistus running into another, or veins of schistus running into granite, as well as of granite running into schistus. He will find it a hard task to restrain the activity of his theory, and to confine its explanations to those things that really exist.

302. As the Huttonian system cannot boast of theories of equal versatility, it would be not a little embarrassed to account for veins of great magnitude proceeding from a rock distinctly stratified, and accompanied with marks of having disturbed the rocks through which they pass. I am, however, inclined to believe, that this embarrassment will never occur; and that the granite veins do not proceed from the rocks that are really stratified, but from such as have never been deposited by water, and where the appearances of stratification, if there are any, are altogether illusory. This anticipation, however, requires to be verified by future observation; and it remains to be seen, whether granitic veins ever accompany real granitic strata, or are peculiar to those in

which the appearances of regular beds are either ambiguous, or are entirely wanting. The decision of this queſtion is an object highly worthy of the attention of geologiſts.

303. An argument, directed at once againſt the igneous origin and unſtratified nature of all granite, is given in a work already mentioned: "If granite had flowed from below, how does it happen, that, after it had burſt through the ſtrata of micaceous ſchiſtus, &c. it did not overflow the neighbouring country? If this hypotheſis were true, Mont Blanc could never have exiſted *."

A theory is never more unfairly dealt with, than when thoſe parts are ſeparated which were meant to ſupport one another, and each left to ſtand or fall by itſelf. This, however, is preciſely what is done in the preſent inſtance; for Dr Hutton's theory of granite would not deſerve a moment's conſideration, if it were ſo inartificially conſtructed, as to ſuppoſe that granite was originally fluid, and yet to point out no means of hindering this fluid from diffuſing itſelf over the ſtrata, and ſettling in a horizontal plane. The truth is, that his theory, at the ſame time that it conceives this ſtone to have been

* Mineralogy of the Scottiſh Iſles, vol. ii. p. 166.

been in fuſion, ſuppoſes it to have been, in that ſtate, injected among the ſtrata already conſolidated; to have heaved them up, and to have been formed in the concavity ſo produced, as in a mould. Thus Mont Blanc, ſuppoſing that it is unſtratified, is underſtood to conſiſt of a maſs that was melted by ſubterraneous heat under the ſtrata, and being impelled upwards by a force, that may ſtand in ſome compariſon with that which projected the planets in their orbits, heaved up the ſtrata by which it was covered, and in which it remained included on all ſides.

304. The covering of ſtrata, thus raiſed up, may have been burſt aſunder at the ſummit, where the curvature and elevation were the greateſt; but the melted maſs underneath may have already acquired ſolidity, or may have been ſuſtained by the beds of ſchiſtus incumbent on its ſides. This ſchiſtus, forming the exterior cruſt, was immediately acted on by the cauſes of waſte and decompoſition, which have long ſince ſtripped the granite of a great part of its covering, and are now exerciſing their power on the central maſs. That even Mont Blanc itſelf, as well as other unſtratified mountains, was once covered with ſchiſtus, will appear to have in it nothing incongruous, when we conſider the height to which the ſchiſtus ſtill riſes on its ſides, or in the adjacent mountains;

 and

and when we reflect, that, from the appearances of waste and degradation which these mountains exhibit, it is certain, that the schistus must have reached much higher than it does at present.

It is obvious, therefore, that when the corresponding parts are brought together, and placed in their natural order, no room is left for the reproach, that this system is inconsistent with the *existence* of granite mountains. I have no pleasure in controversial writing; and, notwithstanding the advantages which a weak attack always gives to a defender, I cannot but regret, that Dr Hutton's adversaries have been so much more eager to refute than to understand his theory.

---

305. A remark which Dr Hutton has made on the quantity of granite that appears at the surface, compared with that of other mineral bodies, has been warmly contested. Having affirmed, that the greater part of rocks bear marks of being formed from the waste and decomposition of other rocks, he alleges that granite, (a stone which does not contain such marks), does not, for as much as appears from actual observation, make up a tenth, nor perhaps even

even a hundredth part of the mineral kingdom *. Mr Kirwan contends, that this is a very erroneous eſtimate, and that the quantity of granite viſible on the ſurface, far exceeds what is here ſuppoſed †. The queſtion is certainly of no material importance to the eſtabliſhment of Dr Hutton's theory: it is evident, too, that an eſtimation, which varies ſo much as from a tenth to a hundredth part, cannot have been meant as any thing preciſe; yet it may not be quite ſuperfluous to ſhow, that the truth probably lies nearer to the leaſt than the greateſt of the limits juſt mentioned.

306. Though granite forms a part, generally the central part, of all the great chains of mountains, it uſually occupies a much leſs extent of ſurface than the primary ſchiſtus. Thus in the Alps, if a line be drawn from Geneva to Ivrea, it will be about eighty-five geographical miles in length, and will meaſure the breadth of this formidable chain of mountains, at the place of its greateſt elevation. Now, from the obſervations of Sauſſure, who croſſed the Alps exactly in this direction, it may be collected, that leſs than nine miles of this line, or not above a tenth part of it, in the immediate vicinity of Mont Blanc, is occupied by granite.

 307. In

* Theory of the Earth, vol. i. p. 211.

† Geol. Eſſays, p. 480.

307. In ſome ſections of the Alps, no granite at all appears. Thus, in the rout from Chambery to Turin, acroſs Mont Cenis, which meaſures by the road not leſs than ninety miles, no granite is found, at leaſt of that kind which is diſtinctly in maſs, and different from gneiſs or veined granite *.

308. In ſome other places of the ſame mountains, the granite is more abundant. A line from the lake of Thun, along the courſe of the Aar, and over the mountains to the upper end of Lago Maggiore, croſſes a very elevated tract, and paſſes by the ſources of the Rhone, the Rhine, and the Teſſino, which laſt runs into the Po. A good deal of granite is diſcovered here, in the mountains of Grimſel and St Gothard; but by far the greater part of it is the veined granite, the granite in maſs being confined chiefly to the north ſide of the Grimſel. Both together do not occupy more than one-third of the line, and therefore the latter leſs than one-ſixth.

309. The eſſay on the mineralogy of the Pyrenees, by the Abbé Palasso, contains a mineralogical chart of thoſe mountains. From this chart I have found, by computation, that the granite does not occupy one-fifth of the horizontal

* Voyages aux Alpes, tom, iii. § 1190, &c.

zontal ſurface on the north ſide of the ridge, reckoning from one end of it to the other. Indeed, many great tracts, even of the central parts of the Pyrenees, contain no granite whatſoever; and not a few of the higheſt mountains conſiſt entirely of calcareous ſchiſtus. A large deduction ſhould be made from the fraction $\frac{1}{5}$, on account of the ſubſtances unknown, which, from the conſtruction of the chart, are often confounded with the granitic tract.

310. I might add other eſtimations of the ſame kind, all confeſſedly rude and imperfect, but ſtill conveying, by means of numbers, a better idea of the limit to which our knowledge approximates, than could be done ſimply by words; and, on the whole, it would appear, that if we ſtate the proportion of granite to ſchiſtus to be that of one to four, we ſhall certainly do no injuſtice to the extent of the former.

It remains to form a rough eſtimate from maps, and from the accounts of travellers, of what proportion of the earth's ſurface conſiſts of primary, and what of ſecondary rocks. After ſupplying the want of accurate meaſurement by what appeared to me the moſt probable ſuppoſitions, I have found, that about $\frac{1}{18}$ of the ſurface of the old continent may be conceived to be occupied by primitive mountains; of which,

if

if we take one-fifth, we have $\frac{1}{90}$ for the part of the ſurface occupied by granite rocks, which differs not greatly from the leaſt of the two limits aſſigned by Dr Hutton.

311. In eſtimating the granite of Scotland, Dr Hutton has certainly erred conſiderably in defect *, and Mr Kirwan, who always differs from him, is here neareſt the truth ; though he is right purely by accident, as the information on which he proceeds is vague and erroneous.

The places in Scotland where granite is found, are very well known ; but the extent of ſome of the moſt conſiderable of them is not accurately aſcertained. In the ſouthern parts, except the granite of Galloway, which is found in two pretty large inſulated tracts, there is no other of any magnitude. The granite of the north extends over a large diſtrict. If we ſuppoſe a line to be drawn, from a few miles ſouth of Aberdeen to a few miles ſouth

* Dr Hutton in this caſe no doubt made a very looſe eſtimate. He ſays, the granite does not perhaps occupy more than a 500dth part of the whole ſurface. The whole ſurface of Scotland is not much more than 23,000 geographical miles, the 500dth part of which is exactly 46 ; and this is exceeded by the granite in Kirkcudbrightſhire alone, as may be gathered from what is ſaid § 282.

ſouth of Fort-William, it will mark out the central chain of the Grampians in its full extent, paſſing over the moſt elevated ground, and by the heads of the largeſt rivers, in Scotland. Along this line there are many granite mountains, and large tracts in which granite is the prevailing rock. There are, however, large ſpaces alſo in which no granite appears, though, if we were permitted to ſpeak theoretically, and if the queſtion did not entirely relate to a matter of obſervation, we might ſuppoſe, that, in no part of this central ridge is the granite far from the ſurface, notwithſtanding that in ſome places it may be covered by the ſchiſtus.

312. A great part of the Grampian mountains is on the ſouth ſide of the line juſt mentioned, but hardly any granite is found in this diviſion of them, except ſuch veins as thoſe of Glentilt. On the north ſide of the line, the granite extends in various directions; and, if from Fort-William a line is drawn to Inverneſs, the quadrilateral figure, bounded on two ſides by theſe lines, and on the other two by the ſea, will be found to contain much granite, and many diſtricts conſiſting entirely of that ſtone. This is in fact the great granite country of Scotland: it is a large tract, containing about 3170 ſquare geographical miles, or about a ſeventh part

part of the whole : but the proportion of it occupied by granite cannot at present be ascertained with any exactness, nor will, till some mineralogist shall find leisure to examine the courses of the great rivers, the Dee, the Spey, &c. which traverse this country. If we call it one-fourth of the whole surface, its extent is certainly not under-rated, and will amount to 790 square miles nearly ; to which adding 150, as a very full allowance for all the other granite contained in Scotland, exclusive of the isles, we shall have 940 square miles, between a twenty-fourth and twenty-fifth part of the surface of the whole.

This computation, it must be observed, aims at nothing precise, but I think it is such, that a more accurate survey would rather diminish than increase the proportion assigned in it to the granite rock.

313. This result may perhaps fall as much short of Mr Kirwan's notion, as it exceeds the estimate made by Dr Hutton. If it shall not, and if the former has, in this instance, come nearest the truth, it cannot be ascribed to the accuracy of his information, or the soundness of the principles which directed his research. Mr WILLIAMS, whom he quotes, was a miner, of great skill and experience in some branches of his profession, to which, if he had confined himself, he might have written a book full of useful

ful information. What he ſays on the ſubject of granite, is, in the main I believe juſt; but it is far too general to authoriſe the concluſion which Mr Kirwan derives from it. Dr Ash, for whoſe judgment I have great reſpect, cannot, I think, have meant, when he uſed the expreſſion granitic rocks, to deſcribe granite ſtrictly ſo called. He ſays, in the paſſage quoted by Mr Kirwan, that "from Galloway, Dumfries, and Berwick, there is a chain of mountains, commonly ſchiſtoſe, but often alſo granitic." Now, the fact is, that the great belt of primary rock, here alluded to, which traverſes the ſouth of Scotland, conſiſts of vertical ſchiſtus of various kinds; but except in Galloway, and again in Lammermuir, near Prieſtlaw, it appears, as already mentioned, to contain no granite whatſoever. If the German mineralogiſt quoted by Mr Kirwan, when he ſays that the Grampian mountains conſiſt of micaceous limeſtone, gneiſs, porphyry, argillite, and granite, alternating with one another, means only to affirm that all theſe ſtones are found in the Grampians, he is certainly in the right, and the catalogue might eaſily be enlarged; but, if he either means to ſay, that theſe are nearly in equal abundance, or that the granite is commonly found in ſtrata alternating with other ſtrata, I muſt ſay, that theſe are propoſitions

poſitions quite contrary to any thing I have ever ſeen or heard of thoſe mountains. But it is probable that this is not meant, and that the fault lies in underſtanding the expreſſions much too literally. Mr Kirwan accuſes Dr Hutton of not knowing where to look for the granite; not aware of how much, notwithſtanding any error committed in the preſent eſtimate, he was ſkilled in the art of mineralogical obſervation; an art, which thoſe who have not practiſed do not always know how to appreciate. But, however imperfect Mr Kirwan's knowledge of this ſubject has been, he has here had the good fortune to correct a mineralogiſt of very ſuperior information. The mere diſpoſition to oppoſe is not always without its uſe: no man is in every thing free from error, and, to controvert indiſcriminately all the opinions of any individual, is an infallible ſecret for being ſometimes in the right.

Note xvi. § 100.

Rivers and Lakes.

314. Rivers are the cauſes of waſte moſt viſible to us, and moſt obviouſly capable of producing

producing great effects. It is not, however, in the greateſt rivers, that the power to change and wear the ſurface of the land is moſt clearly ſeen. It is at the heads of rivers, and in the feeders of the larger ſtreams, where they deſcend over the moſt rapid ſlope, and are moſt ſubject to irregular or temporary increaſe and diminution, that the cauſes which tend to preſerve, and thoſe that tend to change the form of the earth's ſurface, are fartheſt from balancing one another, and where, after every ſeaſon, almoſt after every flood, we perceive ſome change produced, for which no compenſation can be made, and ſomething removed which is never to be replaced. When we trace up rivers and their branches toward their ſource, we come at laſt to rivulets, that run only in time of rain, and that are dry at other ſeaſons. It is there, ſays Dr Hutton, that I would wiſh to carry my reader, that he may be convinced, by his own obſervation, of this great fact, *that the rivers have, in general, hollowed out their valleys.* The changes of the valley of the main river are but ſlow; the plain indeed is waſted in one place, but is repaired in another, and we do not perceive the place from whence the repairing matter has proceeded. That which the ſpectator ſees here, does not therefore immediately ſuggeſt to him what has been the ſtate of things before the valley was hollowed

hollowed out. But it is otherwise in the valley of the rivulet; no person can examine it without seeing, that the rivulet carries away matter which cannot be repaired, except by wearing away some part of the surface of the place upon which the rain that forms the stream is gathered. The remains of a former state are here visible; and we can, without any long chain of reasoning, compare what has been with what is at the present moment. It requires but little study to replace the parts removed, and to see nature at work, resolving the most hard and solid masses, by the continued influences of the sun and atmosphere *. We see the beginning of that long journey, by which heavy bodies travel from the summit of the land to the bottom of the ocean, and we remain convinced, that, *on our continents, there is no spot on which a river may not formerly have run* †.

315. The view thus afforded of the operations, in their nascent state, which have shaped out and fashioned the present surface of the land, is necessary to prepare us for following them to the utmost extent of their effects. From these effects, the truth of the proposition, that rivers have cut and formed, not the

* Theory of the Earth, vol. ii. p. 294.

† *Ibid.* p. 296.

the beds only, but the whole of the valleys, or rather ſyſtem of valleys, through which they flow, is demonſtrated on a principle which has a cloſe affinity to that on which chances are uſually calculated, § 99. In order to conceive rightly the courſe of a great river, and the communication ſubſiſting between the main trunk and its remoteſt branches, let us take the inſtance of the Danube, and caſt our eyes on one of the maps conſtructed by MARSIGLI, for illuſtrating the natural hiſtory of that great river *. When it is conſidered, that over all the vaſt and uneven ſurface, which reaches from the Alps to the Euxine, and from the mountains of Crapack to thoſe of Hæmus, a regular communication is kept up between every point and the line of greateſt depreſſion, in which the river flows, no one can heſitate to acknowledge, that it is the agency of the waters alone which has opened them a free paſſage through all the intricacies of this amazing labyrinth. In effect, ſuppoſe this communication to be interrupted, and that ſome ſudden operation of nature were to erect a barrier of mountains to oppoſe the Theiſe or the Drave, as they rolled their waters to the Danube. From this what could poſſibly reſult, but the damming up of thoſe rivers till

 their

* Hiſtoire du Danube, tom. i. tab. 34.

their waters were deep, or high enough to find a vent, either under the bafes or over the tops of the oppofing ridge. Thus there would be formed immenfe lakes and immenfe cataracts, which, by filling up what was too low, and cutting down what was too high, would in time reftore fuch a uniform declivity of furface as had before prevailed. Juft fo in the times that are paft, whatever may have been the irregularities of the furface at its firft emerging from the fea, or whatever irregularities may have been produced in it by fubfequent convulfions, the flow action of the ftreams would not fail in time to create or renew a fyftem of valleys communicating with one another, like that which we at prefent behold. Water, in all circumftances, would find its way to the loweft point; though, where the furface was quite irregular, it would not do fo till after being dammed up in a thoufand lakes, or dafhed in cataracts over a thoufand precipices. Where neither of thefe is the cafe; and where the lake and the cataract are comparatively rare phenomena; there we perceive that conftitution of a furface, which water alone, of all phyfical agents, has a tendency to produce; and we muft conclude, that the probability of fuch a conftitution having arifen from another caufe, is, to the probability of its having

having arifen from the running of water, in fuch a proportion as unity bears to a number infinitely great.

316. The courfes of many rivers retain marks that they once confifted of a feries of lakes, which have been converted into dry ground, by the twofold operation of filling up the bottoms, and deepening the outlets. This happens, efpecially, when fucceffive terraces of gravelly and flat land are found on the banks of a river, § 100. Such platforms, or *haughs* as they are called in this country, are always proofs of the wafte and *detritus* produced by the river, and of the different levels on which it has run; but they fometimes lead us farther, and make it certain, that the great mafs of gravel which forms the fucceffive terraces on each fide of the river, was depofited in the bafon of a lake. If, from the level of the higheft terrace, down to the prefent bed of the river, all is alluvial, and formed of fand and gravel, it is then evident, that the fpace as low as the river now runs muft have been once occupied by water; at the fame time, it is clear, that water muft have ftood, or flowed as high at leaft, as the uppermoft furface of the meadow. It is impoffible to reconcile thefe two facts, which are both undeniable, but by fuppofing a lake, or body of ftagnant water, to have here occupied a great hollow, (which by us muft be held as one of the original

 nal

nal inequalities of the globe, becaufe we can trace it no farther back), and that this hollow, in the courfe of ages, has been filled up by the gravel and alluvial earth brought down by the river, which is now cutting its channel through materials of its own depofiting. There is no great river that does not afford inftances of this, both in the hilly part of its courfe, and where it defcends firft from thence into the plain. Were there room here for the minuter details of topographical defcription, this might be illuftrated by innumerable examples.

317. It is faid above, that the water muft have run or ftood, in former times, as low as the prefent bottom of the river; but there is often clear evidence, that it has run or ftood much lower, becaufe the alluvial land reaches far below the prefent level of the river. This is known to hold in very many inftances, where it has happened that pits have been funk to confiderable depths on the banks of large rivers. By that means, the depth of the alluvial ground, under the prefent bed of the river, has been difcovered to be great; and from this arifes the difficulty, fo generally experienced, of finding good foundations for bridges that are built over rivers in large vallies, or open plains, the ground being compofed of travelled materials to an unknown depth, without any thing like the native

or

or ſolid ſtrata. In ſuch caſes, it is evident, that formerly the water muſt have been much lower, as well as much higher, than its preſent level, and this is only conſiſtent with the notion, that the place was once occupied by a deep lake.

318. If, following the light derived from theſe indications, we go back to the time when the river ran above the higheſt of thoſe levels at which it has left any traces of its operations, we ſhall ſee it compoſed of a ſeries of lakes and cataracts, from which, by the filling up of the one, and the wearing down of the other, the waters have at length worked out to themſelves a quiet and uninterrupted paſſage to the ocean. We may, indeed, on good evidence, go back ſtill farther than the ſucceſſion of ſuch meadows or terraces, as are above mentioned, will carry us, and may conſider the whole valley, or *trough* of the river, as produced by its own operations. The original inequalities of the ſurface, and the diſpoſition of the ſtrata, muſt no doubt have determined the water-courſes at firſt; but this does not hinder us from conſidering the rivers as having modified and changed thoſe inequalities, and as the *proximate* cauſes of tne ſhape and configuration which the ſurface has now aſſumed.

319. From this gradual change of lakes into rivers, it follows, that a lake is but a temporary and accidental condition of a river, which is

 every

every day approaching to its termination; and the truth of this is atteſted, not only by the lakes that have exiſted, but alſo by thoſe that continue to exiſt. Where any conſiderable ſtream enters a lake, a flat meadow is uſually obſerved increaſing from year to year. The ſoil of this meadow is diſpoſed in horizontal ſtrata: the meadow is terminated by a marſh; which marſh is acquiring ſolidity, and is ſoon to be converted into a meadow, as the meadow will be into an arable field. All this while the ſediment of the river makes its way ſlowly into the lake, forming a mound or bank under the ſurface of the water, with a pretty rapid ſlope toward the lake. This mound increaſes by the addition of new earth, ſand, and gravel, poured in over the ſlope; and thus the progreſs of filling up continually advances.

320. In ſmall lakes, this progreſs may eaſily be traced; and will be found ſingularly conſpicuous in that beautiful aſſemblage of lakes, which ſo highly adorns the mountain ſcenery of Weſtmoreland and Cumberland. Among theſe a great number of inſtances appear, in which lakes are either partially filled up, or have entirely diſappeared. In the Lake of Keſwick, we not only diſcover the marks of filling up at the upper end, which extend far into Borrowdale, from which valley a ſmall river flows into the lake; but we have the cleareſt proof, that this lake was once united

united to that of Baſſenthwaite, and occupied the whole valley from Borrowdale to Ouſe-Bridge. Theſe two lakes are at preſent joined only by a ſtream, which runs from the former into the latter, and their continuity is interrupted by a conſiderable piece of alluvial land, compoſed of beds of earth and gravel, without rock, or any appearance of the native ſtrata. This ſeparation, therefore, ſeems no other than a *bar*, formed by the influx of two rivers, that enter the valley here from oppoſite ſides, the Greata from the eaſt, and Newland's Water from the weſt. The ſurface of this meadow is at preſent twelve or fifteen feet at leaſt above the level of either lake ; and a quantity of water of that depth muſt therefore have been drawn off by the deepening of the iſſue at Ouſe-Bridge, through which the water of both lakes paſſes, in its way to the ocean.

Many more examples, ſimilar to this, may be collected from the ſame lakes ; there are indeed few places from which, in this branch of geology, more information may be collected.

321. The larger lakes exemplify the ſame progreſs. Where the Rhone enters the Lake of Geneva, the beach has been obſerved to receive an annual increaſe ; and the Portus Valeſiæ, now Prevailais, which is at preſent half a league from the lake, was formerly cloſe upon its bank. Indeed, the ſediments of the Rhone appear clearly to

have formed the valley through which it runs, to a diſtance of about three leagues at leaſt from the place where the river now diſcharges itſelf into the lake. The ground there is perfectly horizontal, compoſed of ſand and mud, little raiſed above the level of the river, and full of marſhes. The depoſition made by the Rhone after it enters the lake, is viſible to the eye; and may be ſeen falling down in clouds to the bottom.

The great lakes of North America are undergoing the ſame changes, and, it would ſeem, even with more rapidity. As the rivers, however, which ſupply theſe vaſt reſervoirs, are none of them very great, the filling up is much leſs remarkable than the draining off of the water, by the deepening of the outlet. An intelligent traveller has remarked, that in Lake Superior itſelf the diminution of the waters is apparent, and that marks can be diſcovered on the rocks, of the ſurface having been ſix feet higher than it is at preſent. In the ſmaller lakes this diminution is ſtill more evident *. In ſome of thoſe far inland, the ground all round appeared to the ſame traveller to be the depoſite from the rivers, of which the lakes themſelves may be conſidered as a mere expanſion †.

322. In

* Mackenzie's Voyages through the Continent of North America to the Frozen and Pacific Oceans, p. xlii. and xxxvi.

† *Ibid.* p. 122.

322. In order to give uniform declivities to the rivers, the lakes muſt not only be filled up or drained, but the cataract, wherever there is one, muſt be worn away. The latter is an operation in all caſes viſible. The ſtream, as it precipitates itſelf over the rocks, hurries along with it, not only ſand and gravel, but occaſionally large ſtones, which grind and wear down the rock with a force proportioned to their magnitude and acceleration. The ſmooth ſurface of the rocks in all waterfalls, their rounded ſurface, and curious excavations, are the moſt ſatisfactory proofs of the conſtant attrition which they endure; and, where the rocks are deeply interſected, theſe marks often reach to a great height above the level on which the water now flows. The phenomena, in ſuch inſtances, are among the arguments beſt calculated to remove all incredulity reſpecting the waſte which rivers have produced, and are continuing to produce. They ſuffer no doubt to remain, that the height and aſperity of every waterfall are continually diminiſhing; that innumerable cataracts are entirely obliterated; that thoſe which remain are verging toward the ſame end, and that the Falls of Montmorenci and Niagara muſt ultimately diſappear.

323. Though there can be no doubt of the juſtneſs of the preceding concluſions, when applied to

to lakes in general, ſome apparent exceptions occur, in which the progreſs of draining and filling up ſeems to have been ſuſpended, or even to have gone in a contrary direction. Theſe exceptions conſiſt of the lakes which appear to have received a greater quantity of materials than was ſufficient to have filled them up. Such, for example, is the Lake of Geneva, which receives the Rhone deſcending from the Vallais, one of the deepeſt and longeſt vallies on the ſurface of the earth. Now, if this valley, or even a large proportion of it, had been excavated by the Rhone itſelf, as our theory leads us to ſuppoſe, the lake ought to have been entirely filled up, becauſe the materials brought down by the river ſeem to be much greater than the lake, on any reaſonable ſuppoſition concerning its original magnitude, can poſſibly have received. What, then, it may be ſaid, has become of all that the Rhone has brought down and depoſited in it? The lake, at this moment retains, in ſome places, the depth of more than 1000 feet; and yet, of all that the Rhone carries into it, nothing but the pure water iſſues. If it has been continuing to diminiſh, both in ſuperficial extent and in depth, from the time when the Rhone began to run into it, what muſt have been its original dimenſions?

I cannot pretend to remove entirely the difficulty which is here ftated; yet I think the following remarks may go fome length in doing fo.

324. It is certain, that from the prefent ftate of the lake of Geneva, and of the ground round it, we can hardly draw any inference as to its original dimenfions. Sauffure has traced, with his ufual fkill, the marks of the courfe of the Rhone, on a level greatly above the prefent; and, by obfervations on the fide of Mount Saleve, has found proofs of the running of water, at leaft 200 toifes above the prefent fuperficies of the lake. But, if ever the fuperficies of the lake ftood at this height, or at this height nearly, though we can conjecture but little concerning the ftate of the adjacent country, which no doubt was alfo on a higher level, the lake may very well be fuppofed to have been of far greater dimenfions than it is now. It may have occupied the whole fpace from Jura to Saleve, and included the Lake of Neufchatel; fo that it may have been of magnitude fufficient to receive the fpoils of the Vallais, which, as the furface of its waters lowered, may have been wafhed away and carried down to the fea. Thus it may have afforded a temporary receptacle for the *debris* of the Alps, and may have ferved for an *entrepot*, as it were, where thofe debris were depofited,

depofited, before they were carried to the place of their ultimate deftination.

325. But the great depth which the lake has at prefent, ftill remains to be explained, becaufe no mud or gravel could be carried beyond the gulf, of a thoufand feet deep, which was here ready to receive it. The reality of this difficulty muft be acknowledged; and fome caufe feems to act, if not in the generation, yet certainly in the prefervation of lakes, with which we are but little acquainted. We can indeed imagine fome caufes of that kind to occur in the courfe of the degradation of the land, which may produce new lakes, or increafe the dimenfions of the old. The wearing away of a ftratum, or body of ftrata, may lay bare, and render acceffible to the water, fome beds of mineral fubftances foluble in that fluid. The diftrict, for inftance, in Chefhire, which contains rock-falt, extends over a tract of fourteen or fifteen miles, and is covered by a thick ftratum of clay, more or lefs indurated, which defends the falt from the water at the furface, and preferves the whole mafs in a ftate of drynefs. Should this covering be broke open by any natural convulfion, or fhould it be worn away, as it muft be in the progrefs of the general detritus, the water would gain admiffion to the faline ftrata,

would

would gradually diſſolve them, and form of courſe a very deep and extenſive lake, where all was before dry land. This event is not only poſſible, but it ſhould ſeem, that in the courſe of things it muſt neceſſarily happen.

326. Something of this kind may have taken place in the track of the Rhone, and may have produced the Leman Lake. It is not impoſſible, that, at a very remote period, the Rhone deſcended from the Alps without forming any lake, or at leaſt any lake of which the remains are now exiſting; and this ſuppoſition, which is more probable than that of § 324, we ſhall ſoon find to be conformable to appearances of another kind. The river may have wore away the ſecondary limeſtone ſtrata over which it took its courſe after it left the ſchiſtus of the mountains; and, in doing ſo, may have reached ſome ſtratum of a ſaline nature, and this being waſhed out, may have left behind it a lake, which is but modern compared with many of the revolutions that have happened on the ſurface of the earth *.

This explanation is no doubt hypothetical; but it is propoſed in one of thoſe caſes, in which

* There are ſalt ſprings at Bex, near Aigle, about ten miles from the head of the lake: ſaline ſtrata, therefore, are probably at no great diſtance.

which hypothetical reaſonings are warranted by the ſtricteſt rules of philoſophical inveſtigation. It is propoſed in a caſe, where the cauſes viſible to man ſeem inadequate to the effect, and where we muſt therefore have recourſe to an agent that is inviſible. If the operations aſcribed to this agent are conformable to the analogy of nature, it is all that can in reaſon be required.

327. Another circumſtance may alſo influence the generation and preſervation of lakes; but it is alſo one with which we are but little acquainted. The ſtrata, and indeed the whole body of mineral ſubſtances which forms the baſis of our land, have been raiſed up from the bottom of the ſea, by a progreſs that ſhould ſeem in general to have been gradual and ſlow. Appearances, however, are not wanting, which ſhew, that this progreſs is not uniform; and that both riſing and ſinking in the ſurface of the land, or in the rocks which are the baſe of it, have happened within a period of time, which is by no means of great extent. In this progreſs, the elevations and depreſſions may not be the ſame for every ſpot. They may be partial, and one part of a ſtratum, or body of ſtrata, may riſe to a greater height, or be more depreſſed, than another. It is not impoſſible, that this proceſs may affect the depth of lakes,

lakes, and change the relative level of their ſides and bottom.

328. All lakes, however, do not involve the difficulty which the preceding conjectures are intended to remove. The great lakes of North America do not, for inſtance, receive their ſupply from very large rivers. Of courſe, it is not from a tract great in compariſon of themſelves, that the waſte and detritus is brought down into them; and it ſeems not at all wonderful, that, without being filled up, they have been able to receive it. The ſame, in a degree at leaſt, is true of many other lakes.

It ſhould alſo be conſidered, that we may err greatly in the eſtimate we make of the materials actually carried down and depoſited in any lake. To judge of their entire amount, we ſhould know the original form of the inequalities on the earth's ſurface; of the quantity of depreſſion which exiſted, independently of the rivers; and though, in general, theſe original inequalities may be overlooked, and the preſent conſidered as made by the running of water, yet, in particular inſtances, this may be far from true. The Vallais, for example, which we conſider as the work of the Rhone, may, when the Alps roſe out of the ſea, have included many depreſſions of the ſurface, which the river joined together, and, from being a ſeries of lakes, formed into one great valley.

329. The

329. The mouths by which rivers on bold rocky coafts difcharge their waters into the fea, afford a very ftriking confirmation of the conclufions concerning the general fyftem of wafte and degradation which have been drawn above. At thefe mouths we ufually fee, not only the bed of the river, but frequently a confiderable valley, cut out of the folid rock, while that rock perferves its elevation, and its precipitous afpect, wherever it is not interfected by a run of water. No convulfion that can have torn afunder the rocks; no breach that can have been made in them, antecedent to the running of the waters, will account for the circumftance of every river finding a correfponding opening, by which it makes its way to the fea; for that opening being fo nearly proportional to the magnitude of the river, and for fuch breaches never occurring but where ftreams of water are found.

330. The actual furvey of any bold and rocky coaft, will make this clearer than any general ftatement can poffibly do. Let us take, for an example, the coaft of the Britifh Channel, from Torbay to the Land's End, which is faced by a continued rampart of high cliffs, formed of much indurated and primeval rock. If we confider the breaches in this rampart, at the mouths of

of the Dart, of the Plym and Tamer, of the river at Fowey, of the Fal, the Hel, &c. it will appear perfectly clear, that they have been produced by their respective streams. Where there is no stream, there is no breach in the rock, no softening in the bold and stern aspect which this shore every where presents to the ocean. If we look at the smaller streams, we find them working their way through the cliffs at the present moment; and we see the steps by which the larger valleys of the Dart and the Tamer have been cut down to the level of the sea. If we would have still clearer evidence, that no breaches made antecedently to the running of the rivers have opened a way for them, we need only look to the opposite side, or northern shore, of the same promontory, where we also find a series of outlets, all originating in the ridge of the country, and becoming deeper as they approach the sea, but altogether unconnect d with the openings on the south side; and this could hardly have been the case, had they been the effects of previous concussions, or of any peculiarity in the original structure of the rocks.

331. In contemplating such coasts as these, when we go back to the time when the rivers ran upon a level as high as the highest of the cliffs on the sea-shore, we must suppose, that the land then extended many miles farther into what is

 now

now occupied by the ſea. When at Plymouth, for inſtance, the Tamer and the Plym flowed on the level of Mount Edgecombe or of Staten Heights, if the rivers ran with a moderate declivity into the ſea, the coaſt muſt have advanced many miles beyond its preſent line. Thus the land, when higher, was alſo more extended, and the limits of our iſland in that ancient ſtate, were doubtleſs very different from theſe by which it is at preſent circumſcribed.

If with the ſame views we conſider any other of the bold coaſts which the map of the world preſents us with, we ſhall quickly remark, that wherever a deep interſection of the ſea is made into the land, as on the weſtern ſhores of our own iſland, or on thoſe of Norway, a river runs in at the head of it, and points out by what means ſuch inlets are formed, viz. by the united powers of the ſea and of the land, the waters of the latter having opened the way by which thoſe of the former have penetrated ſo far into the country.

332. It is not meant aſſuredly to deny the irregularities of the ſea-coaſt, as it may have originally exiſted; theſe irregularities no doubt determined the initial operations of that waſte and decay, by which, in proceſs of time, they were themſelves entirely effaced. The line of our

our coafts may be compared to one of thofe curves, which are fometimes treated of in the higher geometry, where the ordinates are functions, not only of their abfciffæ, but alfo of the time elapfed fince a certain epocha. The form of the curve at that epocha, or when the time began to flow, correfponds to the original form of the fea-coaft, on its emerging from the ocean, and before the powers of wafting and decay had begun to act upon it. To fpeak ftrictly, the original figure, in both cafes, influences all the fubfequent; but the farther removed from it in point of time, the lefs is that influence; fo that, in phyfical queftions, and for the purpofe of fuch approximations as fuit the imperfection of our knowledge, the confideration of the original figure may be wholly left out.

NOTE XVII. § 105.

Remains of Decompofed Rocks.

333. THE plain of Crau was the *Campus Lapideus* of the ancients; and, as mythology always feeks to connect itfelf with the extraordinary facts in natural hiftory, it was faid to be the fpot where Hercules, fighting with the fons

of

of Neptune, and being in want of weapons, was ſupplied from heaven by a ſhower of ſtones: hence it was called *Campus Herculeus*.

This plain is on the eaſt ſide of the Rhone, between Salon and Arles: it is of a triangular form, about twenty ſquare leagues in extent, and is covered almoſt entirely with quartzy gravel. This immenſe collection of gravel has been ſuppoſed by ſome to have been brought down by the Durance from the Alps of Dauphiny; by others it has been aſcribed to the Rhone; and by many to the ſea, as being a work too great for any river. The explanation mentioned above, § 105, namely, that the looſe gravel on the plain ariſes from the decompoſition of a great ſtratum of puddingſtone, which is the baſis of the whole, is the opinion of Sauſſure, and is founded on his own obſervations *.

334. The theories that have been contrived for explaining the phenomena of the plain of Crau, afford an inſtance of the neceſſity of generalizing our obſervations before we can explain a particular appearance: in other words, they prove the

* See Voyages aux Alpes, tom. iii. § 1592, et 1597. See alſo on this ſubject a Memoir by Lamanon, Journal de Phyſique, tom. xxii. p. 477; and another by M. De Servieres, *ibid.* p. 270.

the truth of Lord Bacon's maxim, That the explanation of a phenomenon ſhould not be ſought for from the ſtudy of that phenomenon alone, but from the compariſon of it with others. One of the theories of this plain is, that the breccia, which is the baſe of it, is formed from the conſolidation of the looſe gravel of the plain, by water percolating through it, and carrying ſome cementing ſubſtance along with it, or ſome *lapidific juice*, as it is called. And indeed, whether the gravel is formed from the breccia, or the breccia from the gravel, is a queſtion which probably could never be reſolved by the mere examination of the plain itſelf. But the queſtion is very ſoon decided, when we compare what is obſerved here with other appearances in the natural hiſtory of the earth's ſurface, and conſider how much more frequent the decompoſition of ſolids is, than their reconſolidation, in any place above the level of the ſea.

335. The argument for the decompoſition of ſtony ſubſtances which is afforded by the ſtate of this ſingular plain, may be confirmed by the appearances obſerved in many extenſive tracts of land all over the world, and eſpecially in ſome parts of Great Britain. The road to Exeter from Taunton Dean, between the latter and Honiton, paſſes over a large heath or down, conſiderably elevated above the plain of Taunton. The rock which

which is the baſe of this heath, as far as can be diſcovered, is limeſtone, and over the ſurface of it large flints, in the form of gravel, are very thickly ſpread. There is no higher ground in the neighbourhood from which this gravel can be ſuppoſed to have come, nor any ſtream that can have carried it, ſo that no explanation of it remains, but that it is formed of the flints contained in beds of limeſtone, which are now worn away. The flints on the heath are preciſely of the kind found in limeſtone; many of them are not much worn, and cannot have travelled far from the rock in which they were originally contained. It ſeems certain, therefore, that they are the *debris* of limeſtone ſtrata, now entirely decompoſed, that once lay above the ſtrata which at preſent form the baſe of this elevated plain, and probably covered them to a conſiderable height. This explanation carries the greater probability with it, that any other way of accounting for the fact in queſtion, as the travelling of the gravel from higher grounds, or the immerſion of the ſurface under the ſea, will imply changes in the face of the country, incomparably greater than are here ſuppoſed. Our hypotheſis ſeems to give the *minimum* of all the kinds of change that can poſſibly account for the phenomenon.

336. The

336. The ſame remarks may be made on the high plain of Blackdown, which the road paſſes over in going from Exeter to the weſtward. The flints there are diſſeminated over the ſurface as thickly as in the other inſtance, and can be explained only on the ſame ſuppoſition.

Again, in the interior of England, beginning from about Worceſter and Birmingham, and proceeding north-eaſt through Warwickſhire, Leiceſterſhire, Nottinghamſhire, as far as the ſouth of Yorkſhire, a particular ſpecies of highly indurated gravel, formed of granulated quartz, is found every where in great abundance. This ſame gravel extends to the weſt and north-weſt, as far as Aſhburn in Derbyſhire, and perhaps ſtill farther to the north. The quantity of it about Birmingham is very remarkable, as well as in many other places; and the phenomenon is the more ſurpriſing, that no rock of the ſame ſort is ſeen in its native place. It is ſuch gravel as might be expected in a mountainous country, in Scotland, for inſtance, or in Switzerland, but not at all in the fertile and ſecondary plains of England.

This enigma is explained, however, when it is obſerved, that the baſis of the whole tract juſt deſcribed is a red ſandſtone, often containing in it a hard quartzy gravel, perfectly ſimilar to that which has juſt been mentioned. From the diſſolution of beds of this ſandſtone, which formerly covered the preſent, there can be no doubt that this gravel is derived. But, as the gravel is in general thinly diſperſed through the ſandſtone, and abounds only in ſome of its layers, it ſhould therefore ſeem, that a vaſt body of ſtrata muſt have been worn away and decompoſed, before ſuch quantities of gravel as now exiſt in the ſoil could have been let looſe.

337. I have ſaid, that a rock capable of affording ſuch gravel as this, is not to be found in the tract of country juſt mentioned. This, however, is not ſtrictly true; for in Worceſterſhire, between Bromeſgrove and Birmingham, about ſeven miles from the latter, a rock is found conſiſting of indurated ſtrata, greatly elevated, and without doubt primitive, from the detritus of which ſuch gravel as we are now ſpeaking of might be produced. Theſe ſtrata ſeem to riſe up from under the ſecondary, where they are interſected by the road; and, for as much as appears, are not of great thickneſs, ſo that they cannot have afforded the materials of this gravel directly, though they may have done ſo indirectly, or through the medium of the red ſandſtone; that is to ſay, a primary rock of which they are the remains, may have afforded materials for the gravel in the ſandſtone; and this ſandſtone may in its turn have afforded the materials

terials of the prefent foil, and particularly the gravel contained in it.

338. Pudding-ftones being very liable to decompofition, have probably, in moft countries, afforded a large proportion of the loofe gravel now found in the foil. The mountains, or at leaft hills, of this rock, which are found in many places, prove the great extent of fuch decompofition. Mount Rigi, for inftance, on the fide of the Lake of Lucerne, is entirely of pudding-ftone, and is 742 toifes in height, meafured from the level of the lake. By the defcriptions given of it, as well as of other hills of the fame kind in Switzerland, we may, without due attention, be led to fuppofe that they are entirely formed of loofe gravel. Even M. Sauffure's defcription is chargeable with this fault, though, when attended to, it will be found to contain a fufficient proof, that this hill is compofed of real pudding-ftone *. The nature of the thing alfo, would be fufficient to convince us, that a hill, more than 4000 feet in height, could not confift of loofe and unconfolidated materials.

If, then, we regard Mount Rigi as the remains of a body of pudding-ftone ftrata, we muft conclude, that thefe ftrata were originally more extenfive, and the adjacent valleys and plains will ferve,

* Voyages aux Alpes, tom, iv. § 1941.

ferve, in fome degree, to meafure the quantity of them which time has deftroyed.

339. If the theory of unftratified mountains, namely thofe of whinftone, porphyry, and granite, be admitted as laid down above, it will furnifh a meafure of the deftruction which has taken place in the ftratified rocks, and of the vaft depredations which have been made upon them fince they were raifed up from the bottom of the fea. Like every other meafure, however, of wafting, by a thing that is itfelf fubject to wafte, it can only give a *minimum*, or a limit which the quantity wafted muft neceffarily exceed.

The abrupt face of a whinftone rock muft be underftood as an evidence, that fome body of ftrata which fupported it when fluid, remained in contact with it, when it was become folid; and if this part of the mould in which the whinftone was caft, has difappeared, it muft generally be afcribed to the operation of wafte and decompofition. Such a face, for inftance, as that which Salifbury *Craig* prefents to the weft, viz. a perpendicular wall of whinftone, about ninety feet high, raifed on a body of fandftone ftrata of the height of about 300 feet, can have been produced only by having been abutted againft fome ftratified rock, equally abrupt, and

and of the ſame elevation with itſelf. Of this rock no part remains.

The baſaltic rock of Edinburgh Caſtle is nearly in the ſame ſtate. Its perpendicular ſides on the ſouth, weſt, and north, are now diſengaged from the ſtrata by which they were once encompaſſed.

340. The granite mountains alſo, where they are quite unſtratified, give riſe to the ſame concluſion. Thoſe central chains which we find in ſo many inſtances towering above the ſchiſtus which cover their ſides, have probably been once completely enveloped by the latter; and, on this ſuppoſition, an eſtimate may ſometimes be formed of the original height of ſuch mountains.

In theſe eſtimations, however, ſome uncertainty muſt ariſe, from our being unable to diſtinguiſh between the effects which are to be aſcribed to the fracture and diſlocation that took place when the compound body of ſtratified and unſtratified rocks was raiſed up from the bottom of the ſea, and the effects produced by the ſubſequent waſte and decompoſition at the ſurface. In this, as in many other inſtances, we are not always able to ſeparate between the original inequalities of the ſurface, and thoſe which wearing has produced.

341. It would be important to aſcertain the rate at which the elevation of mountains decreaſes, and this

this is what we may perhaps expect to be accomplished, by the progress of geological science, and the multiplying of accurate observations. It has been supposed, that the Pyrenees diminish about ten inches in a century; but what confidence is to be put in this estimate, I am unable to determine *.

A very unequivocal mark of the degradation of mountains is often to be met with in the heaps of loose stones found on their tops. These stones, it is obvious, cannot have come from any other place by natural means, and they are accordingly always sharp and angular, and have none of the characters of transported rocks. They are said sometimes to have been brought by men's hands; but this is highly improbable, their quantity is often so considerable, and the difficulty of transportation so great. Where any purpose was to be served by heaping them together, men have availed themselves of the stones that they found ready prepared on the summit, and have constructed from them cairns, which have served as signals, useful in their pastoral, and sometimes in their military occupations.

NOTE

* Essai sur le Mineralogie des Pyrenées, p. 87.

NOTE XVIII. § 112.

Tranſportation of Stones, &c.

342. NATURE ſupplies the means of tracing with conſiderable certainty the migration of foſſil bodies on the ſurface of the earth, as only the more indurated ſtones, and thoſe moſt ſtrongly characteriſed, can endure the accidents that muſt befal them in travelling to a diſtance from their native place.

It is a fact very generally obſerved, that where the valleys among primitive mountains open into large plains, the gravel of thoſe plains conſiſts of ſtones, evidently derived from the mountains. The nearer that any ſpot is to the mountains, the larger are the gravel ſtones, and the leſs rounded is their figure; and, as the diſtance increaſes, this gravel, which often forms a ſtratum nearly level, is covered with a thicker bed of earth or vegetable ſoil. This progreſſion has particularly been obſerved in the valleys of Piedmont and the plains of Lombardy, where a bed of gravel forms the baſis of the ſoil, from the foot of the Alps to the ſhores of the Hadriatic.

Hadriatic *. We may collect from GUETTARD, that a ſimilar gradation is found in the gravel and earth which cover the great plain of Poland, from Mount Krapack to the Baltic †. The reaſon of this gradation is evident; the farther the ſtones have travelled, and the more rubbing they have endured, the ſmaller they grow, the more regular is the figure they aſſume, and the greater the quantity of that finer detritus which conſtitutes the ſoil. The waſhing of the rains and rivers is here obvious; and each of the three quantities juſt mentioned, if not directly proportional to the diſtance which the ſtones have migrated from their native place, may be ſaid, in the language of geometry, to be at leaſt proportional to a certain function of that diſtance.

343. The immenſe quantity of *cailloux roulés*, or rounded gravel, collected in the immediate vicinity of mountainous tracts, has led ſome geologiſts to ſuppoſe the exiſtence of ancient currents, which deſcended from the mountains, in a quantity, and with a *momentum*, of which there is no example in the preſent ſtate of the world. Thus Sauſſure imagines, that the hill of Supergue, near Turin, which is formed of gravel, can only be explained by ſuppoſing ſuch currents

* Voyages aux Alpes, tom. iii. § 1315.

† Mém. Acad. des Sciences, 1762, p. 234; 293, &c.

rents as are juſt mentioned, or what he terms a *debacle*, to have taken place at ſome former period*. If, however, we aſcribe to the mountains a magnitude and elevation vaſtly greater than that which they now poſſeſs; if we regard the vallies between them as cut out by the rivers and torrents from an immenſe rampart of ſolid rock, neither materials ſufficiently great, nor agents ſufficiently powerful, will appear to be wanting, for collecting bodies of gravel and other looſe materials, equal to any that are found on the ſurface of the earth. The neceſſity of introducing a *debacle*, or any other unknown agent, to account for the tranſportation of foſſils, ſeems to ariſe from under-rating the effects of action long continued, and not limited by ſuch ſhort periods as circumſcribe the works, and even the obſervations, of men.

344. The ſupply of gravel and *cailloux roulés*, for the plains extended at the feet of primitive mountains, is doubtleſs in many caſes much increaſed by the pudding-ſtone, interpoſed between the ſecondary and the primary ſtrata. The beds of pudding-ſtone contain gravel already formed on the ſhores of continents, that ceaſed to exiſt before the preſent were produced; and the cement of this

* Voyages aux Alpes, tom. iii. § 1303.

this gravel, yielding eaſily to the weather, allows the ſtones included in it to be waſhed down by the torrents, and ſcattered over the plains. I know not if the hill of Supergue, above mentioned, is not in reality a maſs of the pudding-ſtone which forms the border of the Alps, and of which the materials have ſuffered no tranſportation ſince the time of their laſt conſolidation. This at leaſt is certain, that Sauſſure, notwithſtanding his accuracy, has ſometimes confounded the looſe gravel on the ſurface with that which is conſolidated into rock; an inaccuracy which is to be charged, as I have elſewhere obſerved, rather againſt his ſyſtem than himſelf.

345. The looſe ſtones found on the ſides of hills, and the bottoms of valleys, when traced back to their original place, point out with demonſtrative evidence the great changes which have happened ſince the commencement of their journey; and in particular ſerve to ſhow, that many valleys which now deeply interſect the ſurface, had not begun to be cut out when theſe ſtones were firſt detached from their native rocks. We know, for inſtance, that ſtones under the influence of ſuch forces as we are now conſidering, cannot have firſt deſcended from one ridge, and then aſcended on the ſide of an oppoſite ridge. But the granite of

of Mont Blanc has been found, as mentioned above, on the ſides of Jura, and even on the ſide of it fartheſt from the Alps. Now, in the preſent ſtate of the earth's ſurface, between the central chain of the Alps, from which theſe pieces of granite muſt have come, and the ridge of Mont Jura, beſides many ſmaller valleys, there is the great valley of the Rhone, from the bottom of which, to the place where they now lie, is a height of not leſs than 3000 feet. Stones could not, by any force that we know of, be made to aſcend over this height. We muſt therefore ſuppoſe, that when they travelled from Mont Blanc to Jura, this deep valley did not exiſt, but that ſuch a uniform declivity, as water can run on with rapidity, extended from the one ſummit to the other. This ſuppoſition accords well with what has been already ſaid concerning the recent formation of the Leman Lake, and of the preſent valley of the Rhone.

346. We can derive, in a matter of this ſort, but little aid from calculation; yet we may diſcover by it, whether our hypotheſis tranſgreſſes materially againſt the laws of probability, and is inconſiſtent with phyſical principles already eſtabliſhed. The horizontal diſtance from Mont Jura to the granite mountains, at the head of the Arve, may be accounted fifty geo-

 graphic

graphic miles. Though we ſuppoſe Mont Blanc, and the reſt of thoſe mountains, to have been originally much higher than they are at preſent, the ridge of Jura muſt have been ſo likewiſe; and though probably not by an equal quantity, yet it is the faireſt way to ſuppoſe the difference of their height to have been nearly the ſame in former ages that it is at preſent, and it may therefore be taken at 10,000 feet. The declivity of a plane from the top of Mont Jura to the top of Mont Blanc, would therefore be about one mile and three quarters in fifty, or one foot in thirty; an inclination much greater than is neceſſary for water to run on, even with extreme rapidity, and more than ſufficient to enable a river or a torrent to carry with it ſtones or fragments of rock, almoſt to any diſtance.

Sauſſure, in relating the fact that pieces of granite are found among the high paſſes near the ſummits of Mont Jura, alleges, that they are only found in ſpots from which the central chain of the Alps may be ſeen. But it ſhould ſeem that this coincidence is accidental, becauſe, from whatever cauſe the tranſportation of theſe blocks has proceeded, the form of the mountains, eſpecially of Mont Jura, muſt be too much changed to admit of the ſuppoſition, that the places of it from which Mont Blanc is now viſible,

viſible, are the ſame from which that mountain was viſible when theſe ſtones were tranſported hither. It may be, however, that the paſſes which now exiſt in Mont Jura are the remains of valleys or beds of torrents, which once flowed weſtward from the lps; and it is atural, that the fragments from the latt r mountains ſhou d be found in the neighbourhood of thoſe ancient water-tracks.

347. Sauſſure obſerved in another part of the Alps, that where the Drance deſcends from the ſides of Mont Velan and the Great St Bernard, to join the Rhone in the Vallais, the valley it runs in lies between mountains of primary ſchiſtus, in which no granite appears, and yet that the bottom of this valley, toward its lower extremity, is for a conſiderable way covered with looſe blocks of granite *. His familiar acquaintance with all the rocks of thoſe mountains, led him immediately to ſuſpect, that theſe ſtones came from the granite chain of Mont Blanc, which is weſtward of the Drance, and conſiderably higher than the intervening mountains. This conjecture was verified by the obſervations of one of his friends, who found the ſtones in queſtion to agree exactly with a rock

* Voyages aux Alpes, tom. ii. § 1022.

rock at the point of Ornex, the nearest part of the granite chain.

In the present state of the surface, however, the valley of Orsiere lies between the rocks of Ornex and the valley of the Drance, and would certainly have intercepted the granite blocks in their way from the one of these points to the other, if it had existed at the time when they were passing over that tract. The valley of Orsiere, therefore, was not formed, when the torrents, or the glaciers transported these fragments from their native place.

Mountainous countries, when carefully examined, afford so many facts similar to the preceding, that we should never have done were we to enumerate all the instances in which they occur. They lead to conclusions of great use, if we would compare the machinery which nature actually employs in the transportation of rocks, with the largest fragments of rock which appear to have been removed, at some former period, from their native place.

348. For the moving of large masses of rock, the most powerful engines without doubt which nature employs are the glaciers, those lakes or rivers of ice which are formed in the highest valleys of the Alps, and other mountains of the first order. These great masses are in perpetual motion,

motion, undermined by the influx of heat from the earth, and impelled down the declivities on which they reſt by their own enormous weight, together with that of the innumerable fragments of rock with which they are loaded. Theſe fragments they gradually tranſport to their utmoſt boundaries, where a formidable wall aſcertains the magnitude, and atteſts the force, of the great engine by which it was erected. The immenſe quantity and ſize of the rocks thus tranſported, have been remarked with aſtoniſhment by every obſerver *, and explain ſufficiently how fragments of rock may be put in motion, even where there is but little declivity, and where the actual ſurface of the ground is conſiderably uneven. In this manner, before the valleys were cut out in the form they now are, and when the mountains were ſtill more elevated, huge fragments of rock may have been carried to a great diſtance; and it is not wonderful, if theſe ſame maſſes, greatly diminiſhed in ſize, and reduced to gravel or ſand, have reached the ſhores, or even the bottom, of the ocean.

349. Next in force to the glaciers, the torrents are the moſt powerful inſtruments employed in

 the

* The ſtones collected on the *Glacier de Miage*, when Sauſſure viſited it, were in ſuch quantity as to conceal the ice entirely. Voyages aux Alpes, tom. ii, § 854.

the tranſportation of ſtones. Theſe, when they deſcend from the ſides of mountains, and even where the declivity of their courſe is not very great, produce effects which nothing but direct experience could render credible. The fragments of rock which oppoſe the torrent, are rendered ſpecifically lighter by the fluid in which they are immerſed, and loſe by that means at leaſt a third part of their weight: they are, at the ſame time, impelled by a force proportional to the ſquare of the velocity with which the water ruſhes againſt them, and proportional alſo to the quantity of gravel and ſtones which it has already put in motion. Perhaps, after taking all theſe circumſtances into computation, in the midſt of a ſcene perfectly quiet and undiſturbed, a philoſopher might remain in doubt as to the power of torrents to move the enormous bodies of rock which are ſeen in the bottom of the narrow valleys or deep glens of a mountainous country; but his incredulity, ſays an experienced traveller, will ceaſe altogether, if he has been ſurpriſed by a ſtorm in the midſt of ſome Alpine region; if he has ſeen the number and impetuoſity of the cataracts which ruſhed down the ſides of the mountains, and beheld the ruin which accompanied them; and if, when the tempeſt was paſſed, he has viewed thoſe meadows,

which

which a few hours before were covered with verdure, now buried under heaps of ſtones, or overwhelmed by maſſes of liquid mud, and the ſides of the mountains cut by deep ravines, where the track of the ſmalleſt rivulet was not before to be diſcovered *.

It is but rarely, however, even on occaſions like theſe, that ſuch vaſt maſſes of rock can be ſeen actually in motion, as are often found on the ſurface, apparently removed to a great diſtance from their native place. The magnitude of theſe is ſo great, in many inſtances, that their tranſportation cannot be explained without ſuppoſing, that the ſurface was very different when theſe tranſportations took place from what it is at preſent; that the elevation of the mountains was greater, and the ground ſmoother and more uniform, at leaſt in ſome directions. If theſe ſuppoſitions are admitted, and they are countenanced, as we have already ſeen, by almoſt every phenomenon in geology, the difficulties which preſent themſelves here will not appear inſurmountable.

350. One of the largeſt blocks of granite that we know of, is on the eaſt ſide of the lake of

 Geneva,

* See an account of a thunder ſtorm near Bareges, in the Eſſai ſur la Mineralogie des Pyrenées, p. 134.

Geneva, called *Pierre de Gouté*, about ten feet in height, with a horizontal ſection of fifteen by twenty *. Another block not far from it, and nearly of the ſame ſize, has ſome remains of ſchiſtus attached to it. Theſe ſtones very much reſemble thoſe which have fallen from the *Aiguilles*, in the valley of Chamouny. The diſtance from their preſent ſituation to thoſe *Aiguilles* is about thirty Engliſh miles, with many mountains and valleys at preſent interpoſed. By whatever means, therefore, theſe blocks were tranſported, their motion muſt have been over a ſurface of much more uniform declivity than the preſent. If the ſurface was without great inequalities, and its general declivity about one foot in thirty, as already computed, the glaciers, in the firſt place, and the torrents afterwards, may have ſerved for the tranſportation even of theſe rocks.

351. Again, in the narrow vale or glen which ſeparates the Great from the Little Saleve, the ſtrata are all calcareous, but a great number of looſe blocks of granite and primary ſchiſtus are ſcattered over the ſurface. A block of the former, near the lower end of the valley, is about the ſize of 1200 cubic feet. Two other large blocks of the ſame kind of ſtone reſt on a baſe of horizontal limeſtone,

* Voyages aux Alpes, tom. i. § 308.

limeſtone, elevated two or three feet above the reſt of the ſurface. This elevation ariſes no doubt from the protection which the ſtones have afforded to the calcareous beds on which they lie, ſo that theſe beds do not wear away ſo faſt as thoſe which are fully expoſed to the weather. But it is ſurely to take a very limited view of the operations on the ſurface, to ſuppoſe, with Sauſſure, that the parts of the calcareous rock under theſe ſtones has ſuffered no waſte whatſoever, ſo that the ſtones remain now in the identical ſpot where they were placed by the great *debacle* which brought them down from the high Alps *. For my part, I have no doubt that the Arve, which is ſtill at no great diſtance, when it ran on a higher level, and in a line different from the preſent, aided by the glaciers and ſuperior elevation of the mountains, was an engine ſufficiently powerful for effecting the tranſportation of theſe ſtones.

352. Theſe phenomena are not peculiar to the Alps, but prevail, in a greater or leſs degree, in the vicinity of all primary or granite mountains. In the iſland of Arran, a fragment of the ſame kind with that which conſtitutes the upper part of Goatfield, is found on the ſea-ſhore, at leaſt three miles from the neareſt granite rock, and with

* *Ibid.* § 227.

with a bay of the ſea intervening. Its dimenſions are not far from thoſe of the *pierre de gouté.* In ſome former ſtate of the granitic mountains in that iſland, the declivity from the top of Goatfield may have been very uniform, and more rapid than it is at preſent.

353. Beſides glaciers and torrents, which have no doubt been the principal inſtruments in producing theſe changes, other cauſes may have occaſionally operated. Large ſtones, when once detached, and reſting on an inclined plane, from the effects of waſte and decompoſition, may advance horizontally, at the ſame time that they deſcend perpendicularly, and this will happen though they be not urged by any torrent, or any thing but their own weight; for the ſurface of the ground, as it waſtes, remains higher under the ſtone, and for a little way round it, than at a greater diſtance, on account of the protection which it receives from the ſtone, as in the inſtances at Saleve, juſt mentioned. The ſtone itſelf alſo becomes rounded at the bottom; and thus the ſurface in contact with the ground is diminiſhed in extent, and the two ſurfaces rendered convex towards one another. It muſt therefore happen, that the ſupport, continually weakening, will at length give way, and the ſtone incline or roll toward the lower ſide, and may even roll conſiderably,

ſiderably, if its centre of gravity has been high above its point of ſupport, and if its ſurface has had much convexity : Thus the horizontal may very far exceed the perpendicular motion ; and, in the courſe of ages, the ſtone may travel to a great diſtance. A ſtone, however, which travels in this manner, muſt diminiſh as it proceeds, and muſt have been much greater in the beginning than it is at preſent.

354. This kind of motion may be aided by particular circumſtances. When a ſtone reſts on an inclined plane, ſo as to be in a ſtate not very remote from equilibrium, if a part be taken away from the upper ſide, the equilibrium will be loſt, and the ſtone will thereby be put in motion. That ſtones which lie on other ſtones, may, by wearing, be brought very near an equilibrium, is proved by what are called *rocking-ſtones*, or in Cornwall *Logan ſtones*, which have ſometimes been miſtaken for works of art ; but that are certainly nothing elſe than ſtones, which have been ſubjected to the univerſal law of waſting and decay, in ſuch peculiar circumſtances, as nearly to bring about an equilibrium of that ſtable kind, which, when ſlightly diſturbed, re-eſtabliſhes itſelf *. The logan ſtone at the

* I do not preſume ſo far as to ſay, that all rocking-ſtones are produced by natural means : I have not ſuffi-cient

the Land's End, is a maſs of granite, weighing more than ſixty tons, reſting on a rock of granite, of conſiderable height, and cloſe on the ſea-ſhore. The two ſtones touch but in a ſmall ſpot, their ſurfaces being conſiderably convex towards one another. The uppermoſt is ſo nearly in an equilibrium, that it can be made to vibrate by the ſtrength of a man, though to overſet it entirely would require a vaſt force. This ariſes from the centre of gravity of the ſtone being ſomewhat lower than the centre of curvature of that part of it on which it has a tendency to roll; the conſequence of which is, that any motion impreſſed on the ſtone, forces its centre of gravity to riſe, (though not very conſiderably), by which means it returns whenever the force is removed, and vibrates backward and forward, till it is reduced to reſt. Were it required to remove the ſtone from its place, it might

---

cient information to juſtify that aſſertion; but the great ſize of that at the Land's End, its elevated poſition, and the approaches toward ſomething of the ſame kind which are to be ſeen in other parts of that ſhore, prove that it is no work of art. They who aſcribe it to the Druids, do not conſider the rapidity with which the Corniſh granite waſtes, nor think how improbable it is, that the conditions neceſſary to a rocking-ſtone, whether produced by nature or art, ſhould have remained the ſame for ſixteen or ſeventeen hundred years.

might be moſt eaſily done, by cutting off a part from one ſide, or blowing it away by gunpowder; the ſtone would then loſe its balance, would tumble from its pedeſtal, and might roll to a conſiderable diſtance. Now, what art is here ſuppoſed to perform, nature herſelf in time will probably effect. If the waſte on one ſide of this great maſs ſhall exceed that on the oppoſite in more than a certain proportion, and it is not likely that that proportion will be always maintained, the equilibrium of the Logan ſtone will be ſubverted, never to return. Thus we perceive how motion may be produced by the combined action of the decompoſition and gravitation of large maſſes of rock.

355. Beſides the gradual waſte to which ſtones expoſed to the atmoſphere are neceſſarily ſubject, thoſe of a great ſize appear to be liable to ſplitting, and dividing into large portions, no doubt from their weight. This may be obſerved in almoſt all ſtones that happen to be in ſuch circumſtances as we are now conſidering; and from this cauſe the ſubverſion of their balance may be more ſudden, and of greater amount, than could be expected from their gradual decay.

Thus, if to the waſting of a ſtone at the bottom, we add the accidents that may befal it in the waſting of its ſides, we ſee at leaſt the phyſical poſſibility of detached ſtones being put in motion,

motion, merely by their own weight. It is indeed remarkable, that ſome of the largeſt of theſe ſtones reſt on very narrow baſes. Thoſe at the foot of Saleve touch the ground only in a few points: The Boulder-ſtone of Borrowdale is ſupported on a narrow ridge like the keel of a ſhip, and is prevented from tumbling by a ſtone or two, that ſerve as a kind of ſhores to prop it up. Very unexpected accidents ſometimes happen to diſturb the reſt of ſuch fragments of rock as have once migrated from their own place. Sauſſure mentions a great maſs of *lapis ollaris* *, that lies detached on the ſide of a declivity in the valley of Urſeren, in the canton of Uri. The people uſe this ſtone as a quarry, and are working it away on the upper ſide, in conſequence of which it will probably be ſoon overſet, and will roll to the bottom of the valley.

356. In many inſtances it cannot be doubted, that ſtones of the kind here referred to are the remains of maſſes or veins of whinſtone or granite, now worn away, and that they have travelled but a very ſhort way, or perhaps not at all, from their original place. Many of the large blocks of whinſtone which we find in this country, ſometimes ſingle, and ſometimes ſcattered

---

* Voyages aux Alpes, tom. iv. § 1851.

tered in confiderable abundance over a particular fpot, are certainly to be referred to this caufe. But the moft remarkable examples of this fort are the ftones found at the Cape of Good Hope, on the hill called *Paarlberg*, which takes its name from a chain of large round ftones, like the pearls of a necklace, that paffes over the fummit. Two of thefe, placed near the higheft point, are called the Pearl and the Diamond, and were mentioned feveral years ago in the Philofophical Tranfactions *. From a more recent account, thefe ftones appear to be a fpecies of granite, though the hill on which they lie is compofed of fandftone ftrata †. The Pearl is a naked rock, that rifes to the height of 400 feet above the fummit of the hill; the Diamond is higher, but its bafe is lefs, and it is more inacceffible.

From the above ftones forming a regular chain, as well as from the immenfe fize of the two largeft, it is impoffible to fuppofe that they have been moved; and it is infinitely more probable, that they are parts of a granite vein, which runs acrofs the fandftone ftrata, and of which fome parts have refifted the action of the weather, while the reft have yielded to it. The whole

* Vol. lxviii. p. 102.

† Barrow's Travels into Southern Africa, p. 60.

whole geological hiſtory of this part of Africa ſeems highly intereſting, ſince, as far as can be collected from the accounts of the ingenious traveller juſt mentioned, it conſiſts of horizontal beds of ſandſtone or limeſtone, reſting immediately on granite, or on primary ſchiſtus. Looſe blocks of granite are ſeen in great abundance at the foot of the Table Mountain, and along the ſea-ſhore.

---

357. The ſyſtem which accounts for ſuch phenomena as have been conſidered in this and ſome of the preceding notes, by the operation of a great deluge, or *debacle* as it is called, has been already mentioned. In Dr Hutton's theory, nothing whatever is aſcribed to ſuch accidental and unknown cauſes; and, though their exiſtence is not abſolutely denied, their effects, whatever they may have been, are alleged to be entirely obliterated, ſo that they can be referred to no other claſs but that of mere poſſibilities. A minute diſcuſſion, however, of the queſtion, Whether there are, on the ſurface of the earth, any effects that require the interpoſition of an extraordinary cauſe, would lead into a longer digreſſion than is ſuited to this place. I ſhall briefly ſtate what appear to be the principal

cipal objections to all such explanations of the phenomena of geology.

358. The general structure of valleys among mountains, is highly unfavourable to the notion that they were produced by any single great torrent, which swept over the surface of the earth. In some instances, valleys diverge, as it were from a centre, in all directions. In others, they originate from a ridge, and proceed with equal depth and extent on both sides of it, plainly indicating, that the force which produced them was *nothing*, or evanescent at the summit of that ridge, and increased on both sides, as the distance from the ridge increased. The working of water collected from the rains and the snows, and seeking its way from a higher to a lower level, is the only cause we know of, which is subject to this law.

359. Again, if we consider a valley as a space, which perhaps with many windings and irregularities, has been hollowed out of the solid rock, it is plain, that no force of water, suddenly applied, could loosen and remove the great mass of stone which has actually disappeared. The greatest column of water that could be brought to act against such a mass, whatever be the velocity we ascribe to it, could not break asunder and displace beds of rock many leagues in length, and in continuity with

the rock on either ſide of them. The ſlow working of water, on the other hand, or the powers that we ſee every day in action, are quite ſufficient for this effect, if time only is allowed them.

360. Some valleys are ſo particularly conſtructed, as to carry with them a ſtill ſtronger refutation of the exiſtence of a *debacle*. Theſe are the longitudinal valleys, which have the openings by which the water is diſcharged, not at one extremity, but at the broadſide. Such is that on the eaſt ſide of Mont Blanc, deeply excavated on the confines of the granite and ſchiſtus rock, and extending parallel to the beds of the latter, from the Col de la Segne to the Col de Ferret; its opening is nearly in the middle, from which the Dorea iſſues, and takes its courſe through a great valley, nearly at right angles to the chain of the Alps, and to the valley juſt mentioned. From the ſtructure of theſe valleys, Sauſſure has argued very juſtly againſt Buffon's hypotheſis, concerning the formation of valleys by currents at the bottom of the ſea *. It affords indeed a complete refutation of that hypotheſis; and it affords one no leſs complete of the ſyſtem which Sauſſure himſelf ſeems on ſome occaſions ſo much inclined to ſupport. For if it be ſaid, that this valley was cut out by the current

* Voyages aux Alpes, tom. ii. § 920.

current of a *debacle*, that current muſt either have run in the direction of the valley of Ferret, or in that of the Dorea, which iſſues from it. If it had the direction of the firſt, it could not cut out the ſecond; and if it had the direction of the ſecond, it could not cut out the firſt. Beſides, the force which excavated this valley muſt have been *nothing* at the two extreme points, viz. at the Col de Segne and the Col de Ferret, and muſt have increaſed with the diſtance from each. It can have been produced, therefore, only by the running of two ſtreams in oppoſite directions, on a ſurface that was but ſlightly uneven, theſe ſtreams at meeting taking a new direction, nearly at right angles to the former. A clearer proof could hardly be required than is afforded in this caſe, that what is now a deep valley was formerly ſolid rock, which the running of the waters has gradually worn away; and that the waters, when they began to run, were on a level as high, at leaſt, as the tops of thoſe mountains by which the valley is bounded toward the lower ſide.

361. Longitudinal valleys, with the water burſting out tranſverſely from their ſides, like the preceding, are by no means confined to mountains of the firſt order. We have a very good example, though on a ſmall ſcale, of a valley of this ſort, within a few miles of Edinburgh.

burgh. The Pentland Hills form a double ridge, ſeparated by a ſmall longitudinal valley, that runs from N. E. to S. W., the water of which iſſues from an opening almoſt in the middle, and directed towards the ſouth. This, therefore, is not the work of any great torrent, which overwhelmed the country; for no one direction, which it is poſſible to aſſign to ſuch a torrent, will afford an explanation, both of the valley and its outlet *.

362. They

* In Scotland there is one valley, of a kind that I believe is extremely rare in any part of the world, in accounting for which, the hypotheſis of a torrent or *debacle* might, if any where, be employed to advantage. This is the valley which extends acroſs the iſland, from Inverneſs to Fort-William, or from ſea to ſea, being open at both ends, and very little elevated in the middle. It is nearly ſtraight, and of a very uniform breadth, except that towards each end it widens conſiderably. The bottom, reckoning tranſverſely, is flat, without any gradual ſlope from the ſides towards the middle. From the ſides the mountains riſe immediately, and form two continued ridges of great height, like ramparts or embankments on each ſide of a large foſſé. A great part of the bottom of this ſingular valley is occupied by lakes, namely, Loch Neſs, Loch Oich, and Loch Lochy. Its length is about ſixty-two miles, and the point of partition from which the waters run different ways, viz. north-eaſt to the German

362. They who maintain the exiſtence of the *debacle*, will no doubt allege, that though theſe

 valleys

German Ocean, and ſouth-weſt to the Atlantic, is between Loch Oich and Loch Lochy; and, by the eſtimation of the eye, I ſhould hardly think that it is elevated more than ten or fifteen feet above the ſurface of either lake. The country on both ſides is rugged and mountainous, and the ſtreams which deſcend from thence into the valley, either fall directly into the lakes, or turn off almoſt at right angles when they enter the valley. Though the bottom of this valley, therefore, is every where alluvial, with the exception, perhaps, of a few rocks which appear at the ſurface, it is certainly not excavated by the rivers which now flow in it. The direction of the valley, it is to be obſerved, is the ſame with that of the vertical ſtrata which compoſe the mountains on either ſide.

Here, then, we have a valley, not cut out by the working of any ſtreams which now appear; and we may therefore make trial of the hypotheſis of a *debacle*. This, however, will afford us no aſſiſtance; becauſe, if we ſuppoſe what is now hollow to have been once occupied by the ſame kind of rock which is on either ſide, no force of torrents can have ſuddenly looſened and removed from its place a body of ſuch vaſt magnitude. A greater column of water, than one having for its baſe a tranſverſe ſection of the valley, could not act againſt it, and this would have to overcome the

valleys were not cut out by means of it, yet others may. But it muſt be recollected, that if ſome

---

the coheſion and inertia of a column of rock of the ſame ſection, and of the length of ſixty-two miles. It is not hazarding much to affirm, that no velocity which could be communicated to water, not even that which it could acquire by falling from an infinite height, could give to it a force in any degree adequate to this great effect.

The explanation of this valley, which appears to me the moſt probable, is the following. It will be ſhewn hereafter, that there is good reaſon to ſuppoſe, that, in moſt parts of our iſland, the relative level of the ſea and land has been in paſt ages conſiderably higher than it is at preſent. In ſuch circumſtances, this valley may have been under the ſurface of the ſea, the higheſt part of it being ſcarcely 100 feet above that level at preſent. It may have been a kind of ſound, therefore, or ſtrait, which connected the German Sea with the Atlantic; and the ſtrong currents, which, on account of the different times of high water in theſe two ſeas, muſt have run alternately up and down this ſtrait, may have produced that flatneſs of the bottom, and ſtraightneſs of the ſides, and that widening at the extremities, which are mentioned above. In this way, too, ſome difficulties are removed relative to Loch Neſs, which is ſo deep as hardly to be conſiſtent with the indefinite length of the period of waſte that muſt be aſcribed to the mountains on each ſide of it. Its depth is ſaid, where greateſt, not to

ſome of the greateſt and deepeſt valleys on the face of the earth, ſuch as that juſt mentioned, on the eaſt ſide of Mont Blanc, are thus ſhewn to be the work of the daily waſting of the ſurface, what other inequalities can be great enough to require the interpoſition of a more powerful cauſe? If a *dignus vindice nodus* does not exiſt here, in what part of the natural hiſtory of the earth is it likely to be found?

363. The large maſſes of rock ſo often met with at a diſtance from their original place, are one of the arguments uſed for the *debacle*. It has, however, been ſhewn, that, ſuppoſing a form of the earth's ſurface conſiderably different from the preſent, eſpecially, ſuppoſing the abſence of the valleys which the rivers have gradually cut out, the tranſportation of ſuch ſtones is not impoſſible, even by ſuch powers as nature employs at preſent. Now, without the ſuppoſition that the ſurface was more continuous, and that its preſent inequalities did not exiſt, no force of torrents, whatever their velocity and magnitude may have been, could have produced this tranſportation. No force of water could raiſe a ſtone like the *pierre de goutté* from the bottom of a valley,

to be leſs than 18c fathoms. According to this hypotheſis, it may, at no very diſtant period, have been a part of the bottom of the ſea.

valley, to the top of a ſteep hill. Indeed, if we ſuppoſe a great fragment of rock to be hurried along on a horizontal or an inclined plane, by the force of water, the moment it comes to a deep valley, and has to riſe up over an aſcent of a certain ſteepneſs, it will remain at reſt; the water itſelf will loſe its velocity, and the heavy bodies which it carried with it will proceed no farther. Thus, therefore, we have the following dilemma. If the ſurface is not ſuppoſed to have had a certain degree of uniformity in paſt times, a *debacle* is inſufficient for the tranſportation of ſtones: If it is ſuppoſed to have had that uniformity, a *debacle* is unneceſſary.

364. Another fact, which has been ſuppoſed favourable to the opinion of the action of great torrents at ſome former period, is, that in countries like that round Edinburgh, where whinſtone hills riſe up from among ſecondary ſtrata, a remarkable uniformity is obſerved in the direction of their abrupt faces. Thus, in the country juſt mentioned, the ſteep faces generally front the weſt, while, in the oppoſite direction, the ſlope is gentle, and the hills decline gradually into the plain. Hence it is ſuppoſed, that a torrent, ſweeping from weſt to eaſt, has carried off the ſtrata from the weſt ſide of theſe hills, but, being obſtructed by the whinſtone rock,

has

has left the ſtrata on the eaſt ſide in their natural place.

But, beſides that no force which can ever be aſcribed to a torrent could have removed at once bodies of ſtrata 300 or 400 feet, nay even 800 or 1000 in thickneſs, which muſt have been the caſe if this were the true explanation of the fact, there is a circumſtance which may perhaps enable us to explain theſe phenomena without the aſſiſtance of any extraordinary cauſe. The ſecondary ſtrata in which the whinſtone hills are found in this part of Scotland, are not horizontal, but riſe or *head* towards the weſt, dipping towards the eaſt. The ſide, therefore, of the whinſtone hills which is precipitous, is the ſame with that towards which the ſtrata riſe. Now, from the manner in which theſe hills are ſuppoſed to have been elevated, the ſtrata are likely to have been moſt broken and ſhattered towards that ſide, while, on the oppoſite, they had the ſupport of the whinſtone rock. They would become a prey, therefore, more eaſily to the common cauſes of eroſion and waſte on the upper ſide than on the lower. The ſtreams that flowed from the higher grounds would wear them on the former moſt readily; and the action of theſe ſtreams would be reſiſted by the ſuperior hardneſs of the whinſtone, juſt as the great torrent of the debacle is ſuppoſed to have been.

It

It ſhould alſo be obſerved, that this fact of the uniform direction of the abrupt faces of mountains, is often too haſtily generalized. In primitive countries, it is no farther obſerved than by the ſteep faces of the mountains being moſt frequently turned toward the central chain. In Scotland, as ſoon as you leave the flat country, and enter the Highlands, the ſcarps of the hills face indiſcriminately all the points of the compaſs, and are directed as often to the eaſt as to the weſt.

365. Where the ſtrata are nearly horizontal, they afford the moſt diſtinct information concerning the direction and progreſs of the waſting of the land. The inclined poſition of the ſtrata, which in all other caſes muſt enter for ſo much into our eſtimate of the cauſes which have produced the preſent inequality of the earth's ſurface, diſappears there entirely; and the whole of that inequality is to be aſcribed to the operations at the ſurface, whether they have been ſudden or gradual. A very important fact from a country of this ſort, is related by Barrow, in his Travels into Southern Africa. The mountains about the Cape of Good Hope, and as far to the north as that ingenious traveller proſecuted his journey, are chiefly of horizontal ſtrata of ſandſtone and limeſtone, exhibiting the appearance, on their abrupt ſides, of regular layers of maſonry, of towers, fortifications, &c. Now, among

among all theſe mountains, he obſerved, that the high or ſteep ſides look conſtantly down the rivers, while the ſloping or inclined ſides have juſt the oppoſite direction. When, in travelling northward, he paſſed the line of partition, where the waters from running ſouth take their direction to the north, he found, that the gradual ſlope, which had hitherto been turned to the north, was now turned to the ſouth: The abrupt aſpect of the mountains, in like manner, from facing the ſouth, was directed to the north; ſo that, in both caſes, the hills turned their backs on the line of greateſt elevation *.

It is evident, therefore, that the form of this land has been determined by the ſlow working of the ſtreams. The cauſes which produced the effects here deſcribed, began their action from the line of greateſt elevation, and extended it from thence on both ſides, in oppoſite directions. This is the moſt preciſe character that can mark the alluvial operations, and diſtinguiſh them from the overwhelming power of a great *debacle*.

366. Laſtly, If there were any where a hill, or any large maſs compoſed of broken and ſhapeleſs ſtones, thrown together like rubbiſh, and neither worked into gravel nor diſpoſed with any regularity, we muſt aſcribe it to ſome other cauſe than

* Barrow's Travels into Southern Africa, p. 245.

than the ordinary *detritus* and wasting of the land. This, however, has never yet occurred; and it seems best to wait till the phenomenon is observed, before we seek for the explanation of it.

367. These arguments appear to me conclusive against the necessity of supposing the action of sudden and irregular causes on the surface of the earth. In this, however, I am perhaps deceived: neither Pallas, nor Saussure, nor Dolomieu, nor any other author who has espoused the hypothesis of such causes, has explained his notions with any precision; on the contrary, they have all spoken with such reserve and mystery, as seemed to betray the weakness, but may have concealed the strength of their cause. I have therefore been combating an enemy, that was in some respects unknown; and I may have supposed him dislodged, only because I could not penetrate to his strong-holds. The question, however, is likely soon to assume a more determinate form. A zealous friend of Dr Hutton's theory, has lately * declared his approbation of the hypothesis which has here been represented as so adverse to that theory; and, from his ability and vigour of research, it is likely to receive every improvement of which it is susceptible.

NOTE

* Transf. Royal Society Edin. vol. v. p. 68.

NOTE XIX. § 117.

Tranſportation of Materials by the Sea.

368. THE exiſtence of the great and extenſive operations, by which the ſpoils of the land are carried all over the ocean, and ſpread out on the bottom of it, may be ſuppoſed to require ſome further elucidation. We muſt attend, therefore, to the following circumſtances.

When the detritus of the land is delivered by the rivers into the ſea, the heavieſt parts are depoſited firſt, and the lighter are carried to a greater diſtance from the ſhore. The accumulation of matter which would be made in this manner on the coaſt, is prevented by the farther operation of the tides and currents, in conſequence of which the ſubſtances depoſited continue to be worn away, and are gradually removed further from the land. The reality of this operation is certain; for otherwiſe we ſhould have on the ſea-ſhore a conſtant and unlimited accumulation of ſand and gravel, which, being perpetually brought down from the land, would continually increaſe on the ſhore, if nature did not employ ſome machinery for removing the advanced

advanced part into the ſea, in proportion to the ſupply from behind.

The conſtant agitation of the waters, and the declivity of the bottom, are no doubt the cauſes of this gradual and widely extended depoſition. A ſoft maſs of alluvial depoſite, having its pores filled with water, and being ſubject to the vibrations of a ſuperincumbent fluid, will yield to the preſſure of that fluid on the ſide of the leaſt reſiſtance, that is, on the ſide toward the ſea, and thus will be gradually extended more and more over the bottom. This will happen not only to the finer parts of the detritus, but even to the groſſer, ſuch as ſand and gravel. For ſuppoſe that a body of gravel reſts on a plane ſomewhat inclined, at the ſame time that it is covered with water to a conſiderable depth, that water being ſubject not only to moderate reciprocations, but alſo to ſuch violent agitation as we ſee occaſionally communicated to the waters of the ocean; the gravel, being rendered lighter by its immerſion in the water, and on that account more moveable, will, when the undulations are conſiderable, be alternately heaved up and let down again. Now, at each time that it is heaved up, however ſmall the ſpace may be, it muſt be ſomewhat accelerated in its deſcent, and will hardly ſettle on the ſame point where it reſted before. Thus it will gain a lit-

tle

tle ground at each undulation, and will flowly make its way towards the depths of the ocean, or to the loweft fituation it can reach. This, as far as we may prefume to follow a progrefs which is not the fubject of immediate obfervation, is one of the great means by which loofe materials of every kind are tranfported to a great diftance, and fpread out in beds at the bottom of the ocean.

369. The lighter parts are more eafily carried to great diftances, being actually fufpended in the water, by which they are very gradually and flowly depofited. A remarkable proof of this is furnifhed from an obfervation made by Lord Mulgrave, in his voyage to the North Pole. In the latitude of 65° nearly, and about 250 miles diftant from the neareft land, which was the coaft of Norway, he founded with a line of 683 fathoms, or 4098 feet; and the lead, when it ftruck the ground, funk in a foft blue clay to the depth of 10 feet*. The tenuity and finenefs of the mud, which allowed the lead to fink fo deep into it, muft have refulted from a depofition of the lighter kinds of earth, which being fufpended in the water, had been carried to a great diftance, and were now without doubt forming

* Phipps's Voyage, p. 74, 141.

forming a regular ſtratum at the bottom of the ſea.

370. The quantity of detritus brought down by the rivers, and diſtributed in this manner over the bottom of the ſea, is ſo great, that ſeveral narrow ſeas have been thereby rendered ſenſibly ſhallower. The Baltic has been computed to decreaſe in depth at the rate of forty inches in a hundred years. The Yellow Sea, which is a large gulf contained between the coaſt of China and the peninſula of Corea, receives ſo much mud from the great rivers that run into it, that it takes its colour, as well as its name, from that circumſtance; and the European mariners, who have lately navigated it, obſerved, that the mud was drawn up by the ſhips, ſo as to be viſible in their wake to a conſiderable diſtance *. Computations have been made of the time that it will require to fill up this gulf, and to withdraw it entirely from the dominion of the ocean: but the data are not ſufficiently exact to afford any preciſe reſult, and are no doubt particularly defective from this cauſe, that much of the earth carried into the gulf by the rivers, muſt be carried out of it by the currents and tides, and the finer parts wafted probably to great diſtances

* Staunton's Account of the Embaſſy to China, vol. i. p. 448.

ſtances in the Pacific Ocean *. The mere attempt, however, towards ſuch a computation, ſhews how evident the progreſs of filling up is to every attentive obſerver ; and, though it may not aſcertain the meaſure, it ſufficiently declares the reality of the operations, by which the waſte of the preſent continents is made ſubſervient to the formation of new land.

371. Sand-banks, ſuch as abound in the German Ocean, to whatever they owe their origin, are certainly modified, and their form determined, by the tides and currents. Without the operation of theſe laſt, banks of looſe ſand and mud could hardly preſerve their form, and remain interſected by many narrow channels. The formation of the banks on the coaſt of Holland, and even of the Dogger Bank itſelf, has been aſcribed to the meeting of tides, by which a ſtate of tranquillity is produced in the waters, and of conſequence a more copious depoſition of their mud. Even the great bank of Newfoundland ſeems to be determined in its extent by the

 action

* Perouſe, in ſailing along the coaſt of China, from Formoſa to the ſtrait between Corea and Japan, though generally fifty or ſixty leagues from the land, had ſoundings at the depth of forty five fathoms, and ſometimes at that of twenty-two. Atlas du Voyage de la Perouſe, No. 43.

action of the gulf-stream. In the North Sea, the current which sets out of the Baltic, has evidently determined the shape of the sand-banks opposite to the coast of Norway, and produced a circular sweep in them, of which it is impossible to mistake the cause.

In proof of the action here ascribed to the waters of the sea, in transporting materials to an unlimited extent, we may add the well-known observation, that the stones brought up by the lead from the bottom of the sea, are generally round and polished, hardly ever sharp and angular. This could never happen to stones that were not subject to perpetual attrition.

372. Currents are no doubt the great agents in diffusing the detritus of the land over the bottom of the sea. These have been long known to exist; but it is only since the later improvements in navigation, that they have been understood to constitute a system of great permanence, regularity and extent, connected with the trade-winds, and other circumstances in the natural history of the globe. The gulf-stream was many years since observed to transport the water, and the temperature of the tropical regions into the climates of the north; and we are indebted to the researches of Major RENNEL, for the knowledge of a great system of currents, of which it is only a part. That geographer, who is so eminent for

for enriching the details of his ſcience with the moſt intereſting facts in hiſtory or in phyſics, has ſhewn, that along the eaſtern coaſt of Africa, from about the mouth of the Red Sea, a current fifty leagues in breadth ſets continually towards the ſouth-weſt *. It doubles the Cape of Good Hope, runs from thence north-weſt, preſerving on the whole the direction of the coaſt, but reaching ſo far into the ocean, that, about the parallel of St Helena, its breadth exceeds 1000 miles. From thence, as it approaches the line, its direction is more nearly eaſt; and meeting in the parallel of 3° north, with a current which has come along the weſtern coaſt of Africa from the north, the two united ſtretch acroſs the Atlantic, in a line ſomewhat ſouth of weſt, and in a very wide and rapid ſtream. This ſtream meets the American land at Cape St Roque, where it is joined by another coming up along the eaſtern ſhore of that continent, and directed towards the north. They proceed northward together till they enter the Gulf of Florida, from which being as it were reflected, they form the Gulf-ſtream, paſſing along the coaſt of North America, and ſtretching acroſs the Atlantic to the Britiſh Iſles. From thence the current turns to the ſouth, and, proceeding down the

* Geography of Herodotus, p. 672.

the coaſt of Spain and Africa, meets the ſtream aſcending from the ſouth, as already deſcribed, and thus continues in perpetual circulation. The velocity of theſe currents is not leſs remarkable than their extent. At the Cape of Good Hope, the rate is thirty nautical miles in twenty-four hours; in ſome places forty-five; and under the line ſeventy-ſeven. When the Gulf-ſtream iſſues from the Straits of Bahama, it runs at the rate of four miles an hour, and proceeds to the diſtance of 1800 miles, before its velocity is reduced to half that quantity. In the parallel of 38°, near 1000 miles from the above ſtrait, the water of the ſtream has been found ten degrees warmer than the air.

373. The courſe of the Gulf-ſtream is ſo fixed and regular, that nuts and plants from the Weſt Indies are annually thrown aſhore on the Weſtern Iſlands of Scotland. The maſt of a man of war, burnt at Jamaica, was driven ſeveral months afterwards on the Hebrides *, after performing a voyage of more than 4000 miles, under the direction of a current, which, in the midſt of the ocean, maintains its courſe as ſteadily as a river does upon the land.

The great ſyſtem of currents thus traced through the Atlantic, has no doubt phenomena correſponding

* Pennant's Arctic Zoology, Introd. p. 70.

correſponding to it in the Indian and Pacific Oceans, which the induſtry of future navigators may diſcover. The whole appears to be connected with the trade-winds, the figure of our continents, the temperature of the ſeas themſelves, and perhaps with ſome inequalities in the ſtructure of the globe. The diſturbance produced by theſe cauſes in the equilibrium of the ſea, probably reaches to the very bottom of it, and gives riſe to thoſe counter currents, which have ſometimes been diſcovered at great depths under the ſurface *.

The great tranſportation of materials that muſt reſult from the action of theſe combined currents is obvious, and ſerves not a little to diminiſh our wonder, at finding the productions of one climate ſo frequently included among the foſſils of another. Amid all the revolutions of the globe, the economy of nature has been uniform, in this reſpect, as well as in ſo many others, and her laws are the only thing that have reſiſted the general movement. The rivers and the rocks, the ſeas and the continents, have been changed in all their parts; but the laws which direct thoſe changes, and the rules to

* Hiſtoire Naturelle de Buffon, ſupplément, tom. ix. p. 479. 8vo.

to which they are ſubject, have remained invariably the ſame.

374. Objections have been made to that tranſlation of materials by the waters of the ocean which is ſuppoſed in this theory, particularly by Mr Kirwan, in his Geological Eſſays; and, though I might perhaps content myſelf with the remark already made, that the Neptunian ſyſtem involves ſuppoſitions concerning the tranſportation of ſolid bodies by the ſea, in the early ages of the world, as wonderful as thoſe which, according to our theory, are common to all ages, I am unwilling to remain ſatisfied with a mere *argumentum ad hominem*, where the fallacy of the reaſoning is ſo eaſily detected.

375. One of Mr Kirwan's objections to the depoſition of materials at the bottom of the ſea, is thus ſtated: "Frisi has remarked, in his mathematical diſcourſes, that if any conſiderable maſs of matter were accumulated in the interior of the ocean, the diurnal motion of the globe would be diſturbed, and conſequently it would be perceptible; a phenomenon, however, of which no hiſtory or tradition gives any account *."

The appeal made here to Friſi is ſingularly unfortunate, as that philoſopher has demonſtrated

* Geol. Eſſays, p. 441.

ted the very contrary of Mr Kirwan's position, and has proved, that the disturbance given to the diurnal motion by the causes here referred to may be real, but cannot be perceptible. Having investigated a formula expressing the law which all such disturbances must necessarily observe, he concludes, " Hâc autem formulâ manifestum fiet, ex iis omnibus variationibus quæ in terrestri superficie observari solent, montium et collium abrasione, dilapsu corporum ponderosiorum in inferiores telluris sinus, nullam oriri posse variationem *sensibilem* diurni motûs. Nam si statuamus data aliqua annorum periodo terrestrem superficiem ad duos usque pedes abradi undique, eam vero materiæ quantitatem ad profunditatem pedum 1000 dilabi; erit omne quod inde orietur incrementum velocitatis diurni motûs $\frac{30000}{(19638051)^2} = \frac{1}{12855068184}$*."

Here, it is evident, that Frisi admits those very changes on the surface which we are contending for, and shews, that their tendency is to accelerate the earth's diurnal motion, but, by a quantity so small, that, in a space of time amounting at least to 200 years, the increase of the diurnal motion would only be such a part of the

 whole

* Frisii Opera, tom. iii. p. 269.

whole as the preceding fraction is of unity *.

376. The

* The time requisite for taking away by waste and erosion two feet from the surface of all our continents, and depositing it at the bottom of the sea, cannot be reckoned less than 200 years. The fraction $\frac{1}{12855068184}$, reduced to parts of a day, is $\frac{1}{148554}$ of a second; so that it would require 200 years to shorten the length of the day, by the above fraction of a second; and therefore it would require 148554 times 200 years, or 29710800 years, to diminish it an entire second. The accumulated effect, however, of all the diminutions during that period, would amount to much more: and if we had any perfectly uniform standard to compare the motion of the earth with, its difference from that standard would increase as the squares of the time, and the total acceleration would amount to one second in 77080 years. Whatever relation this bears to the age of the globe itself, it exceeds more than ten times the age of any historical record.

Though Frisius concludes, as is stated here, that the acceleration produced in the diurnal motion of the earth, is far too inconsiderable to become the object of astronomical observation, he makes a supposition difficult to be reconciled with this conclusion, namely, that the acceleration has had a sensible effect on the figure of the earth, or rather of the sea, having increased the centrifugal force, and thereby accumulated the waters under the equator, in the present, more than in former ages. Such an accumulation, he thinks agreeable to

certain

376. The inſtance juſt given may ſerve as one of many, to ſhew what confidence is to be placed in that indigeſted maſs of facts and quotations which Mr Kirwan, without diſcrimination, and without diſcuſſion, has brought together from all quarters. He has no intention, I believe, to deceive his readers; but we may judge, from this ſpecimen, of the precautions he has taken againſt being deceived himſelf.

In ſome reſpects, the reſult of Friſi's inveſtigation muſt be conſidered as imperfect. If there were no relative motion in the parts of our globe, but that by which things deſcend from a higher to a lower level, a continual acceleration of its rotation, though extremely ſlow, would take place, as above computed. But as, in the interior of the earth, there are undoubtedly motions of a tendency oppoſite to thoſe on the ſurface, and directed from the centre towards the circumference,

---

certain appearances that have been obſerved reſpecting the ancient level of the ſea. Theſe appearances will be afterwards conſidered: it is ſufficient to remark here, that though the fraction, expreſſing the increment of the centrifugal force, muſt be double that which expreſſes the acceleration, it muſt be too ſmall to have any perceptible effect in elevating the ſea, except after an immenſe interval of time; and the compenſations which ariſe from other cauſes, probably muſt prevent it from becoming ſenſible in any length of time whatſoever.

circumference, they muſt produce a retardation in the diurnal revolution; and from this muſt ariſe an inequality, not uniformly progreſſive in the ſame direction, but periodical, and confined within certain limits, as the cauſes are by which it is produced *.

377. Mr

* Even in the deſcent of bodies from a higher to a lower level at the ſurface of the earth, the whole tendency is not to increaſe the velocity of the earth's rotation, and many compenſations take place, which, when the matter is conſidered only in general, are neceſſarily overlooked. This will appear evident, if we reflect, that it is not ſimply the approach of a body towards the centre of the earth, or its removal from that centre, which tends to diſturb the rotation of the earth; but its approach to the axis of the earth, or its removal from that axis. The velocity with which a particle of matter revolves, whether on the ſurface, or in the interior of the globe, is proportional to its diſtance from the axis of rotation; and therefore, when a body comes nearer to the axis, it loſes a part of the motion which it had before; which part, of conſequence, is communicated to the whole maſs of the earth, and therefore tends to increaſe the velocity with which it revolves. The contrary happens when a body recedes from the axis; for it then receives an addition to its velocity, which, of courſe, is taken away from the rotatory motion of the earth.

Hence, bodies moving in a horizontal plane, may increaſe or diminiſh the ſwiftneſs of the diurnal motion, according

377. Mr Kirwan's ſecond objection is founded on the miſapprehenſion of a well-known fact in the

---

according as they move towards the poles or towards the equator; and thoſe which deſcend from a higher to a lower level, diſturb the earth's rotation, much more in conſequence of their horizontal, than of their perpendicular motion. The Ganges, for inſtance, though its ſource is probably elevated no leſs than 7000 feet above the level of the ſea, tends to retard the earth's rotation, by bringing its waters, and the mud contained in them, from the parallel of 31° to that of 22°, and ſo increaſing their diſtance from the earth's axis by more than $\frac{1}{12}$th part. Had the Ganges flowed towards the north, as the Nile does, its effect would have been juſt the contrary.

In the ſame manner, a ſtone deſcending from the top of a mountain, may accelerate or retard the earth's rotation, according to the direction in which it deſcends. If it deſcend on the ſide of the elevated pole, it will then produce acceleration, becauſe its diſtance from the axis will be diminiſhed; but if it deſcend on the ſide of the depreſſed pole, and if the direction in which it is moved, be over a line leſs inclined, than a line drawn from the ſame point to the depreſſed pole, it will then produce a retardation, becauſe its diſtance from the axis will be increaſed.

Let us ſuppoſe, for example, that the top of Mount Blanc is in latitude 45° 49′, and that its height is 2450 toiſes above the level of the ſea. The point at which a line drawn from the top of this mountain, parallel to the

the natural hiſtory of the earth. "Rivers," ſays this author, "do not carry into the ſea the ſpoils which they bring from the land, but employ them in the formation of deltas of low alluvial land at their mouths, according to what Major Rennell has proved." The fact of the formation of *deltas* from the ſpoils which the rivers carry from the

---

the earth's axis, will meet the ſuperficies of the ſea, (ſuppoſing that ſuperficies continued inland from the Mediterranean), muſt be about 2382 toiſes in horizontal diſtance, or about $2\frac{1}{2}$ minutes ſouth of the ſummit, that is, in the parallel of 45° $46\frac{1}{2}'$; and if this parallel be continued all round the globe, the points of the earth's ſurface between it and the equator, are all more diſtant from the earth's axis than the top of Mount Blanc is; whereas all the points to the north of it are nearer to that axis. A ſtone, therefore, from the top of Mount Blanc, if carried any where to the ſouth of the above parallel, will retard the earth's diurnal motion; but if carried any where to the north of the ſame line, will accelerate that motion.

The ſame quantity of matter, however, carried an equal diſtance toward the pole, and toward the equator, from any point, will loſe more velocity in the former caſe than it will gain in the latter, as eaſily follows from the nature of circle. Therefore, ſuppoſing an equal diſperſion of the detritus of a mountain in all directions, the parts that go toward the pole will moſt diſturb the diurnal motion; and hence a balance on their ſide, or in favour of acceleration, as already obſerved.

the higher grounds, is perfectly ascertained; and the detail into which Major Rennel has entered in the passage referred to by Mr Kirwan, does credit to the acuteness and accuracy of that excellent geographer. But it is not there asserted, that rivers employ *all* the materials which they carry with them, in the formation of those deltas, and deliver none of them into the sea. On the contrary, they carry from the *delta* itself mud and earth, which they can deposite nowhere but in the sea; and it is this circumstance chiefly that limits the increase of those alluvial lands, and makes them either cease to increase, or makes them increase very slowly after a certain period, though the supply of earth from the higher grounds remains nearly the same. To make Mr Kirwan's argument conclusive, it would be necessary to prove, that *all* the mud carried down by the Nile or the Ganges, was deposited on the low lands before these rivers enter the sea; a thing so obviously absurd, that nothing but his haste to obtain a conclusion unfavourable to the Plutonic system, could have prevented him from perceiving it *.

378. A

* The instance mentioned in the Geological Essays, from the travels of the Abbé Fortis, concerning urns thrown into the Adriatic, upwards of 1400 years ago, and

378. A remark which Major Rennell has made concerning the mouths of rivers, in his Geography of Herodotus, deſerves Mr Kirwan's attention, though perhaps he may not be able to put on it an interpretation quite ſo favourable to his ſyſtem. The remark is, that the mouths of great rivers are often formed on principles quite oppoſite to one another, ſo that ſome of them have a real delta or triangle of flat land at their mouths, while others have an eſtuary, or what may not improperly be called a *negative* delta. Of the latter kind are ſome of the greateſt rivers in the world, the Plata, the Oroonoko and the Maranon, and by far the greateſt number of our European rivers. Nobody can doubt, that the three rivers juſt named carry with them as much earth as the Nile, or the Euphrates, or any other river in the world. All this they have depoſited in the ſea, and committed to the currents, which ſweep along the ſhore of the American continent, and by theſe they have been ſpread out over the unlimited tracts of the ocean.

Indeed,

---

and not yet covered with mud, muſt be explained from peculiar circumſtances, or local cauſes, with which we are unacquainted. as it makes againſt the depoſition of earth near the ſhore, and in narrow ſeas; a general fact, which, I think, every body admits.

Indeed, nothing can be more juſt than Dr Hutton's obſervation, that where low land is formed at the mouths of rivers, there the rivers bring down more than the ſea is able to carry away; but that where ſuch land is not formed, it is becauſe the ſea is able to carry off immediately all the depoſite which it receives.

379. Mr Kirwan has denied on another principle the power of the ſea to carry to a diſtance the materials delivered into it: "Notwithſtanding," ſays he, "many particles of earth are by rivers conducted to the ſea, yet *none are conveyed to any diſtance*, but are either depoſited at their mouths, or rejected by currents or by tides; and the reaſon is, becauſe the tide of flood is always more impetuous and forcible than the tide of ebb, the advancing waves being preſſed forward by the countleſs number behind them, whereas the retreating are preſſed backward by a far ſmaller number, as muſt be evident to an attentive ſpectator; and hence it is that all floating things caſt into the ſea, are at laſt thrown on ſhore, and not conveyed into the mid regions of the ſea, as they ſhould be if the reciprocal undulations of the tides were equally powerful *."

380. But

* Kirwan's Geol. Eſſays, p. 439

380. But if the *attentive ſpectator*, inſtead of truſting to a vague impreſſion, or liſtening to ſome crude theory of undulations, reflects on one of the moſt ſimple facts reſpecting the ebbing and flowing of the tides, he will be very little diſpoſed to acquieſce in the above concluſion. He has only to conſider, that the flowing of the tide requires juſt ſix hours, and the ebbing of it likewiſe ſix hours; ſo that the ſame body of water flows in upon the ſhore, and retreats from it, in the ſame time. The quantity of matter moved, therefore, and the velocity with which it is moved, are in both caſes the ſame; and it remains for Mr Kirwan to ſhew in what the difference of their force can poſſibly conſiſt.

The force with which the waves uſually break upon our ſhores, does not ariſe from the velocity of the tide being greater in one direction than in another. In the main ocean, the waves have no progreſſive motion, and the columns of water alternately riſe and fall, without any other than a reciprocating motion: a kind of equilibrium takes place among the undulations, and each wave being equally acted upon by thoſe on oppoſite ſides, remains fixed in its place. Near the ſhore this cannot happen; the water on the land ſide from its ſhallowneſs being incapable

of

of rifing to the height neceffary to balance the great undulations which are without. The water runs, therefore, as it were, from a higher to a lower level, fpreading itfelf towards the land fide. This produces the breakers on our fhores, and the furf of the tropical feas. A rock or a fand-bank coming within a certain diftance of the furface, is fufficient, in any part of the ocean, to obftruct the natural fucceffion of undulations; and, by deftroying the mutual reaction of the waves, to give them a progreffive inftead of a reciprocating motion.

381. It is, however, but from a fmall diftance, that the waves are impelled againft the fhore with a progreffive motion. The border of breakers that furrounds any coaft is narrow, compared with the diftance to which the *detritus* from the land is confeffedly carried; the water, while it advances at the furface, flows back at the bottom; and thefe contrary motions are fo nearly equal, that it is but a very momentary accumulation of the water that is ever produced on any fhore.

If it were otherwife, and if it were true that the fea throws out every thing, and carries away nothing, we fhould have a conftant accumulation of earth and fand along all fhores whatfoever, at leaft wherever a ftream ran into the fea.

 This,

This, as is abundantly evident, is quite contrary to the faſt.

So, alſo, the bars formed at the mouths of rivers, after having attained a certain magnitude, increaſe no farther, not becauſe they ceaſe to receive augmentations from the land, but becauſe their diminution from the ſea, increaſing with their magnitude, becomes at length ſo great, as completely to balance thoſe augmentations. When properly examined, therefore, the phenomena, which have been propoſed as moſt inconſiſtent with the indefinite tranſportation of ſtony bodies, afford very ſatisfactory proofs of that operation.

382. It is true, that bodies which float in the water, when carried along on the tops of the waves towards a ſhelving beach, having acquired a certain velocity, are thrown farther in upon the land than the diſtance they would have floated to, if they had been ſimply ſuſtained by the water. The depth of water, therefore, at the place where they take the ground, is not likely to be ſuch as to float them again, and to carry them out towards the ſea. They are, therefore, left behind; and this produces an appearance of a force impelling floating bodies towards the land, much greater and more general than really takes place.

These

Theſe obſervations may ſerve to ſhow, how unſound the principles are from which Mr Kirwan's concluſions are deduced: they are perhaps more than is neceſſary for that purpoſe: it might have been ſufficient to obſerve, that the increaſe of land on the ſea-ſhore is limited, though the augmentation from the land is certainly indefinite, a proof that the diminution from the ſea is conſtant and equal to the increaſe.

383. "Mariners," ſays Mr Kirwan, "were accuſtomed, for ſome centuries back, to diſcover their ſituation, by the kind of earth or ſand brought up by their ſounding plummets; a method which would prove fallacious, if the ſurface of the bottom did not continue invariably the ſame *."

The fact here ſtated, that mariners, when navigation was more imperfect than it is now, had very frequent recourſe to this method, and that they ſtill uſe it occaſionally, is very true. But from this, the only inference that can be fairly deduced is, that the changes at the bottom of the ſea are very ſlow, and the variation but little; not merely from one year to another, but even from one century to another. The rules by which the mariner judged of his poſition from the quality of the earth which the lead brought up, and which were deduced no

 doubt

* Geol. Eſſays, p. 440.

doubt from obſervations made at no very great diſtance of time, might be ſufficient for his purpoſe, though a ſlow change had been all the while going forward. Such obſervations could at beſt have little accuracy, and could not be affected by ſmall variations. It is the ſlowneſs of the change, that makes the experience of one age applicable, in this, as in innumerable other inſtances, to the obſervations of the next. If a long interval is taken, we will look in vain for the ſame uniformity of reſults. A pilot, who would at preſent judge of his poſition in the German Ocean, by comparing his ſoundings with thoſe taken by PYTHEAS, (ſuppoſing them known) in his navigation of that ſea, more than 2000 years ago, could hardly be expected to determine his latitude and longitude with great exactneſs; and I know not if the moſt zealous advocate for the immutability of the earth's ſurface, would be willing to truſt his ſafety in a ſhip that was guided by ſuch antiquated rules.

Note xx. § 118.

Inequalities in the Planetary Motions.

384. The aſſertion that, in the planetary motions, we diſcover no mark, either of the commencement or termination of the preſent order, refers to the late diſcoveries of La Grange and La Place, which have contributed ſo much to the perfection of phyſical aſtronomy. From the principle of univerſal gravitation, theſe mathematicians have demonſtrated, that all the variations in our ſyſtem are periodical; that they are confined within certain limits; and conſiſt of alternate diminution and increaſe. The orbits of the planets change not only their poſition, but even their magnitude and their form: the longer axis of each has a ſlow angular motion; and, though its length remains fixed, the ſhorter axis increaſes and diminiſhes, ſo that the form of the orbit approaches to that of a circle, and recedes from it by turns. In the ſame manner, the obliquity of the ecliptic, and the inclination of the planetary orbits, are ſubject to change; but the changes are ſmall, and, being firſt in one direction, and then in the oppoſite,

they can never accumulate ſo as to produce a permanent or a progreſſive alteration. Thus, in the celeſtial motions, no room is left for the introduction of diſorder ; no irregularity or diſturbance, ariſing from the mutual action of the planets, is permitted to increaſe beyond certain limits, but each of them, in time, affords a correction for itſelf. The general order is conſtant, in the midſt of the variation of the parts ; and, in the language of La Place, there is a certain mean condition, about which our ſyſtem perpetually *oſcillates*, performing ſmall vibrations on each ſide of it, and never receding from it far *. The ſyſtem is thus endowed with a ſtability, which can reſiſt the lapſe of unlimited duration ; it can only periſh by an external cauſe, and by the introduction of laws, of which at preſent no veſtige is to be traced.

385. The ſame *calculus* to which we are indebted for theſe ſublime concluſions, informs us of two circumſtances, which mark the law here treated of as an effect of wiſe deſign, to the entire excluſion both of neceſſity and chance. One of theſe circumſtances conſiſts in the planetary motions being all in the ſame direction, or all *in conſequentia*, as it is called by the aſtronomers.

* Expoſition du Syſtême du Monde, par La Place, Livre iv. chap. 6. p. 199. 2d edit.

mers. This is eſſential to the compenſation and ſtability above mentioned *: had one planet circulated round the ſun in a direction from eaſt to weſt, and another in a direction from weſt to eaſt, the diſturbances they would have produced on one another's motion would not neceſſarily have been periodical; their irregularities might have continually increaſed, and they might have deviated in the courſe of ages from their original condition, beyond any limits that can be aſſigned.

The other circumſtance, on which the ſtability of our ſyſtem depends, is the ſmall eccentricity of the planetary orbits, or their near approach to circles. Were their orbits very eccentric, an opening would be given to progreſſive change, that might ſo far increaſe, as to prove the deſtruction of the whole. But neither the movement of all the planets in the ſame direction, nor the ſmall eccentricity of their orbits, can be aſcribed to accident, ſince that either of theſe ſhould happen by chance, in as many inſtances as there are planets, both primary and ſecondary, is almoſt infinitely improbable. Again, that any neceſſity in the nature of things ſhould have either determined the *direction* of the planetary motions, or proportioned the *quantity* of them to

* La Place, *ibid.*

to the intenſity of the central force, cannot be admitted, as theſe are things unavoidably conceived to be quite independent of one another. It remains, therefore, that we conſider the laws, which make the diſturbances in our ſyſtem correct themſelves, and by that means give firmneſs and permanence to it, as a proof of the conſummate wiſdom with which the whole is conſtructed.

386. The geological ſyſtem of Dr Hutton, reſembles, in many reſpects, that which appears to preſide over the heavenly motions. In both, we perceive continual viciſſitude and change, but confined within certain limits, and never departing far from a certain mean condition, which is ſuch, that, in the lapſe of time, the deviations from it on the one ſide, muſt become juſt equal to the deviations from it on the other. In both, a proviſion is made for duration of unlimited extent, and the lapſe of time has no effect to wear out or deſtroy a machine, conſtructed with ſo much wiſdom. Where the movements are all ſo perfect, their beginning and end muſt be alike inviſible.

NOTE

Note xxi. § 122.

Changes in the apparent Level of the Sea.

387. In ſpeaking of the natural epochas marked out by the phenomena of the mineral kingdom, we have ſuppoſed a greater ſimplicity, and ſeparation of effects from one another, than probably takes place in nature. We have, for inſtance, abſtracted, in ſpeaking of the waſte and degradation of the land, from that elevation which may have been carried on at the ſame time. This appeared neceſſary to be done, in order to ſimplify as much as poſſible the view that was to be given of the whole; but there can be no doubt, that, while the land has been gradually worn down by the operations on its ſurface, it has been raiſed up by the expanſive forces acting from below. There is even reaſon to think, that the elevation has not been uniform, but has been ſubject to a kind of oſcillation, inſomuch, that the continents have both aſcended and deſcended, or have had their level alternately raiſed and depreſſed, independently of all action at the ſurface, and this within

within a period comparatively of no great extent.

It will be eaſily underſtood, that the facts we are going to ſtate, each taken ſingly, prove nothing more than a change of the line in which the ſurface of the ſea interſects the ſurface of the land, leaving it uncertain to which of the two the change ought really to be aſcribed. Taken in combination, however, theſe facts may determine what each of them ſeparately cannot aſcertain. I ſhall firſt, therefore, mention ſome of the principal obſervations relative to the change above mentioned, and ſhall then compare them, in order to diſcover whether it is moſt probable that this change has been produced by the motion of the land or of the ſea.

388. If we begin with examining the coaſts of our own iſland, we ſhall find clear evidence every where, that the ſea once reached higher up upon the land than it does at preſent. The marks of an ancient ſea-beach are to be ſeen beyond the preſent limits of the tide, and beds of ſea-ſhells, not mineralized, are found in the looſe earth or ſoil, ſometimes as high as thirty feet above the preſent level of the ſea. Some of theſe on the ſhores of the Frith of Forth are very well known, and have been often mentioned. Indeed, on the ſhores of that frith, many monuments appear, which would ſeem to carry the difference

difference between the preſent and the ancient level of the ſea, to more than forty feet. The ground on which the Botanic Garden of Edinburgh is ſituated, after a thin covering of ſoil is removed, conſiſts entirely of ſea-ſand, very regularly ſtratified, with layers of a black carbonaceous matter, in thin lamellæ, interpoſed between them. Shells I believe are but rarely found in it, but it has every other appearance of a ſea-beach. The height of this ground above the preſent level of the ſea is certainly not leſs than 40 feet.

389. On almoſt every part of the coaſt where the rocks do not riſe quite abrupt and precipitous from the ſea, ſimilar marks of the lowering of the ſea, or the riſing of the land, may be obſerved. On the ſhores oppoſite to ours, the ſame appearances are remarked. The author of the Lettre Critique to M. de Buffon, tells us, that he had found the bottom of a baſon at Dunkirk, which he had reaſon to think was dug about 950 years ago, ten feet and a half above the preſent low-water mark, though it muſt have been originally under it. The bottom of this baſon is in the native chalk. From this, the ſame author concludes, that the ſea at Dunkirk lowers its level at the rate of an inch nearly in ſeven years. The obſervation was made in

1762,

1762, (Lettre à M. le Comte de Buffon, &c. p. 55.)*.

390. The ſhores of the Low Countries, and of Holland, have been often inſtanced in proof of the ſame kind of changes, and it has been ſuppoſed, that, independently of thoſe artificial barriers which at preſent exclude the waters of the ocean from overflowing a great part of this tract, nature herſelf has brought it nearer to the ſurface than it had formerly been. It is indeed certain, that thoſe countries, to a very great extent inland, have either been under the ſea at ſome period, by no means remote if compared with the great revolutions of the globe, or that they are entirely alluvial, and of the ſame ſort with the Deltas formed at the mouths of rivers. The relative changes, however, of the ſea and land on this tract, have been differently repreſented, and I am unwilling, on

* In the county of Suffolk, near Wood Bridge, at the diſtance of ſeven or eight miles from the ſea, are the Crag-pits, in which prodigious quantities of ſea-ſhells are diſcovered, many of them perfect and quite ſolid, (Pennant's Arctic Zoology, Introd. p. 6.). Lincolnſhire affords various proofs of the ſame kind; but ſome other circumſtances in the appearance of that coaſt, juſt about to be taken notice of, indicate changes of a more complicated nature.

on that account, to found any argument on them.

391. If we proceed farther to the north, to the ſhores of the Baltic for inſtance, we have undoubted evidence of a change of level in the ſame direction as on our own ſhores. The level of this ſea has been repreſented as lowering at ſo great a rate as 40 inches in a century. Celſius obſerved, that ſeveral rocks which are now above water, were not long ago ſunken rocks, and dangerous to navigators; and he particularly took notice of one, which, in the year 1680, was on the ſurface of the water, and in the year 1731 was 20½ Swediſh inches above it. From an inſcription near Aſpô, in the lake Melar, which communicates with the Baltic, engraved, as is ſuppoſed, about five centuries ago, the level of the ſea appears to have ſunk in that time no leſs than 13 Swediſh feet *. All theſe facts, with many more which it is unneceſſary to enumerate, make the gradual depreſſion, not only of the Baltic, but of the whole northern ocean, a matter of certainty.

392. Suppoſing theſe changes of level between the ſea and land to be ſufficiently aſcertained, the ſuppoſition which at firſt occurs is, that the motion

* Friſii Opera, tom. iii. p. 274.

tion has been in the ſea rather than in the land, and that the former has actually deſcended to a lower level. The imagination naturally feels leſs difficulty in conceiving, that an unſtable fluid like the ſea, which changes its level twice every day, has undergone a permanent depreſſion in its ſurface, than that the land, the *terra firma* itſelf, has admi ted of an equal elevation. In all this, however, we are guided much more by fancy than reaſon; for, in order to depreſs or elevate the abſolute level of the ſea, by a given quantity, in any one place, we muſt depreſs or elevate it by the ſame quantity over the whole ſurface of the earth; whereas no ſuch neceſſity exiſts with reſpect to the elevation or depreſſion of the land. To make the ſea ſubſide 30 feet all round the coaſt of Great Britain, it is neceſſary to diſplace a body of water 30 feet deep over the whole ſurface of the ocean. The quantity of matter to be moved in that way is incomparably greater than if the land itſelf were to be elevated; for though it is nearly three times leſs in ſpecific gravity, it is as much greater in bulk, as the ſurface of the ocean is greater than that of this iſland.

393. Beſides, the ſea cannot change its level, without a proportional change in the ſolid bottom on which it reſts. Though there be reaſon to ſuppoſe

poſe that ſuch changes in the bottom do actually take place, yet they are probably much ſlower and more imperceptible than thoſe which we are here conſidering. It is evident, therefore, that the ſimpleſt hypotheſis for explaining thoſe changes of level, is, that they proceed from the motion, upwards or downwards, of the land itſelf, and not from that of the ſea. As no elevation or depreſſion of the ſea can take place, but over the whole, its level cannot be affected by local cauſes, and is probably as little ſubject to variation as any thing to be met with on the ſurface of the globe.

394. Other obſervations, however, made on different ſhores from the preceding, give greater certainty to this concluſion, and make it clear, that the motion or change which we are now treating of is not to be aſcribed to the ſea itſelf.

The obſervations juſt mentioned prove, that the level of the North Sea is lower now than it was heretofore; but it appears, that in the Mediterranean, the oppoſite takes place. Very accurate obſervations made by Manfredi, render it certain, that the ſuperficies of the Hadriatic was higher about the middle of the laſt century, than toward the beginning of the Chriſtian æra.

Some repairs that were carrying on in the cathedral church of Ravenna, in the year 1731, afforded

afforded him an opportunity of obſerving, that the ancient, and probably original, pavement, was four feet and a half below the preſent, and nearly a foot under the level of the ſea at high water *. Now, when the church was built, this cannot have been the poſition of the pavement, relatively to the level of the ſea, for it would have ſubjected the floor to be under water twice in twenty-four hours, and muſt have done ſo the more unavoidably, becauſe at that time (the beginning of the 5th century) the walls of Ravenna were waſhed by the ſea. The fact that this pavement is under the high-water mark, by the quantity juſt mentioned, was aſcertained by actual levelling. This reſult was confirmed by ſimilar facts, obſerved by ZENDRINI at Venice.

395. Manfredi himſelf attributes all this to the elevation of the ſurface of the ſea, and has entered into a long calculation to aſcertain at what rate that ſurface may be ſuppoſed to riſe, on account of the earth and ſand brought down by the rivers, and ſpread out over the bottom of the ſea. But as the fact of the riſe of the level of

---

* Commentarii Academiæ Bononienſis, tom. ii. pars 1ma, p. 237, &c. and pars 2da, p. 1. &c.

of the ſea is not general, and as the contrary is obſerved in the north ſeas, as already proved, this hypotheſis will not explain the apparent riſe in the level of the Hadriatic.

396. Though a local ſubſidence, or ſettling of the ground, could hardly account for this change, the pavement being perfect in its level, and the walls of the cathedral without any ſhake, yet a ſubſidence that has extended to a great tract, as to the whole of Italy, if the maſs moved has continued parallel to itſelf, and changed its place ſlowly, will agree very well with the appearances. The facts here ſtated are alſo the more deſerving of attention, that about Ravenna, the land, at the ſame time that it has ſunk in its level, has extended its ſurface, and has encroached on the ſea. Since the time of AUGUSTUS, the line of the coaſt has been carried farther out by about three miles *. This laſt is the undoubted effect of the degradation of the land by the rivers; and here we have very clear evidence of the forces, both under and above the ſurface, producing their reſpective effects at the ſame time, ſo that while the ſurface is raiſed by earth brought down by the rivers, every given point in

* Manfredi, *ibid*.

the ground is depreſſed and let down to a lower level *.

397. On the ſouthern coaſt of Italy ſimilar facts have been obſerved. BREISLAC, in his *Topographia Fiſica della Campania di Roma* †, from certain appearances in the gulfs of Bajia and Naples, concludes, that at the beginning of the Chriſtian æra, the level of the ſea was lower on that part of the coaſt than it is now. The facts which he mentions are the following: 1*mo*, The remains of an ancient road are now to be ſeen in the Gulf of Bajia at a conſiderable diſtance from the land. 2*do*, Some ancient buildings belonging to Porto Julio are at preſent covered by the ſea. 3*tio*, Ten columns of granite at the foot of Monte Nuovo, which appear to have belonged to the Temple of the Nymphs, are alſo nearly covered by the ſea. 4*to*, The pavement of the Temple of Serapis is now ſomewhat lower than the high-water mark, though it cannot be ſuppoſed that this edifice when built was expoſed to the inconvenience of having its floor frequently under water. 5*to*, The ruins of a palace, built

* On the coaſt of Dalmatia alſo, the riſing of the level of the ſea has been remarked, particularly at the ruins of Diocletian's palace of Spalatro.

† Cap. vi. p. 300.

built by Tiberius in the iſland of Caprea, are now entirely covered by the ſea.

Thus, it appears that the level of the ſea is ſinking in the more northern latitudes, and riſing in the Mediterranean, and it is evident that this cannot happen by the motion of the ſea itſelf. The parts of the ocean all communicating with one another, cannot riſe in one place and fall in another; but, in order to maintain a level ſurface, muſt riſe equally or fall equally over the whole of its extent. If, therefore, we place any confidence in the preceding obſervations, and they are certainly liable to no objection, either from their own nature or the character of the obſervers, we muſt conſider it as demonſtrated, that the relative change of level has proceeded from the elevation or depreſſion of the land itſelf. This agrees well with the preceding theory, which holds, that our continents are ſubject to be acted upon by the expanſive forces of the mineral regions; that by theſe forces they have been actually raiſed up, and are ſuſtained by them in their preſent ſituation.

398. According to ſome other facts ſtated by the ſame ingenious author, it appears, that on the coaſt of Italy the progreſs of the ſea in aſcending, or of the land in deſcending, has not

been uniform during the period above mentioned, but that different oſcillations have taken place ; ſo that, from about the beginning of the Chriſtian æra, till ſome time in the middle ages, the ſea roſe to be ſixteen feet higher than at preſent, from which height it has deſcended till it became lower than it is now, and from that ſtate of depreſſion it is now riſing again. Breiſlac infers this from two facts, which he combines very ingeniouſly with the preceding, viz. the remains of ſome ancient buildings, at the foot of Monte Nuovo, five or ſix feet above the preſent level of the ſea, in which are found the ſhells of ſome of thoſe little marine animals that eat into ſtone : And again, the marble columns of the temple of Serapis, which are alſo perforated by pholades, to the height of ſixteen feet above the ground. All theſe changes Breiſlac aſcribes to the motion of the ſea itſelf ; a ſuppoſition which, as we have ſeen, cannot poſſibly be admitted, ſince nothing can permanently affect the level of the ſea in one place, which does not affect it in all places whatſoever.

399 Appearances, which indicate ſuch alternations as have juſt been mentioned in the level of the ſea, are to be met with on ſome other coaſts. In England, on the coaſt of Lincolnſhire, the remains of a foreſt have been obſerved, which are

now

now entirely covered by the ſea *. The ſubmarine ſtratum which contains the remains of this foreſt, can be traced into the country to a great diſtance, and is found throughout all the fens of Lincolnſhire. The ſtratum itſelf is about four feet thick; it is covered in ſome places by a bed of clay ſixteen feet thick, and under it for twenty feet more is a bed of ſoft mud, like the ſcourings of a ditch, mixed with ſhells and ſilt.

Here then we have a ſtratum which muſt have been once uppermoſt on the ſurface of the dry land, though one part of it is now immerſed under the ſea, and another covered with earth, to the depth of ſixteen feet. A change of level in the ſea itſelf will not explain theſe appearances: they can only be explained by ſuppoſing the whole tract of land to have ſubſided, which is the hypotheſis adopted by the author of the deſcription in the Tranſactions, M. CORRIA DE SERRA; the ſubſidence, however, is not here underſtood to ariſe from the mere yielding of ſome of the ſtrata immediately underneath, but is conceived to be a part of that geological ſyſtem of alternate depreſſion and elevation of the ſurface, which probably extends to the whole mineral kingdom. To reconcile all the differ-

* Phil. Tranſ. 1799. p. 145.

ent facts, I ſhould be tempted to think, that the foreſt which once covered Lincolnſhire, was immerſed under the ſea by the ſubſidence of the land to a great depth, and at a period conſiderably remote; that when ſo immerſed, it was covered over with the bed of clay which now lies on it, by depoſition from the ſea, and the waſhing down of earth from the land; that it has emerged from this great depth till a part of it has became dry land; but that it is now ſinking again, if the tradition of the country deſerves any credit, that the part of it in the ſea is deeper under water at preſent than it was a few years ago. This might alſo ſerve to reconcile, in ſome meaſure, the phenomena of this ſubmarine foreſt with the appearances which indicate an extenſion of the land on the coaſt of Lincolnſhire. Indeed the extenſion of the land is no direct proof, either of its own elevation, or of the depreſſion of the ſea, as we may conclude from the inſtance of Ravenna already mentioned.

400. We have concluded from the facts ſtated above, that the level of the ſea riſes in the Mediterranean, and ſinks in the more northern latitudes; and thence ſome have ſuſpected, that the level of the ſea had in general a tendency to riſe towards the equator, and to ſink towards

wards the poles. This is the notion of Frifi, as has been already remarked, and he fuggefts, that this rife of the fea may be owing to a flight acceleration in the earth's diurnal motion. But there are facts which fhew, that between the tropics the relative level of the fea and land has funk, and is lower at prefent than it was at fome former period, probably not extremely remote. The opinion of Frifi, therefore, is unfupported by obfervation, and, as has been already fhewn, cannot be juftified from theory.

Between the tropics, iflands are formed from the mere accumulation of coral; and it is the peculiarity of thofe regions, to produce rocks that have not paffed through the ufual procefs of mineral confolidation *. The iflots, however, which are thus formed, muft have their bafes laid on a folid rock, though perhaps at a great depth; and it is not probable, that after they are once raifed above the furface of the fea, they can ftill rife farther, except by fome elevation of the rock which ferves as their foundation.

* Dr Fofter, in his Voyage round the World, (vol. ii. p. 146.) gives an inftance in the South Sea Iflands, where the furface of the ifland, though entirely a coral rock, was raifed forty feet above the level of the fea.

tion *. Now, at Palmerſton iſland, which comprehends nine or ten low iſlots, that may be reckoned the heads of a great reef of coral rock, Captain Cook informs us of his having ſeen, " far beyond the reach of the ſea, even in the moſt violent ſtorms, elevated coral rocks, which, on examination, appeared to have been perforated in the ſame manner that the rocks are that now compoſe the outer edge of the reef. This evidently ſhews," he adds, " that the ſea had formerly reached ſo far; and ſome of theſe perforated rocks were almoſt in the centre of the iſland †."

The ſame excellent navigator, giving an account of the peninſula at Cape Denbigh, remarks: " It appeared to me, that this peninſula muſt have been an iſland in remote times; for there were marks of the ſea having flowed over the iſthmus."

401. We are here touching on one of thoſe ſubjects, where we feel much the want of accurate and ancient obſervations, and where it is not from the infancy, but the maturity of ſcience that any thing approaching to certainty can be looked for. The utmoſt that we can expect at preſent, is

* A very curious account of the formation of ſuch iſlands is given by A. Dalrymple Eſq; in the Philoſophical Tranſactions, vol. lvii. p. 394.

† Cook's Third Voyage, vol. i. p. 221.

is an anticipation, which future ages muſt certainly modify, and correct. The beſt thing, in the mean time, that can be done for the advancement of this branch of geological knowledge, is to aſcertain with exactneſs the relative level of the ſea, and of ſuch points upon the land as can be diſtinctly marked, and pointed out to ſucceeding ages. This is not ſo eaſy as it may at firſt appear. Where every object changes, it is difficult to find a meaſure of change, or a fixed point from which the computation may begin. The aſtronomers already feel this inconvenience, and when they would refer their obſervations to an immoveable plane, that ſhall preſerve its poſition the ſame in all ages, they meet with difficulties, which cannot be removed but by a profound mathematical inveſtigation.

In geology, we cannot hope to be delivered from this embarraſſment in the ſame manner; and we have no reſource but to multiply obſervations of the difference of level; to make them as exact as poſſible, and to ſelect points of compariſon that have a chance of being long diſtinguiſhed. The improvements in barometrical meaſurements, which give ſuch facility to the determination of heights, along with ſo conſiderable a degree of accuracy, will furniſh an accumulation of facts that muſt one day be of great value to the geologiſt.

NOTE

Note xxii. § 123.

Fossil Bones.

402. The remains of organised bodies, at present included in the solid parts of the globe, may be divided into three classes. The first consists of the shells, corals, and even bodies of fish, and amphibious animals, which are now converted into stone, and make integrant parts of the solid rock. All these are parts of animals that existed *before the formation of the present land,* or even of the rocks whereof it consists. These remains have been already treated of, and the evidence which they furnish must ever be regarded as of the utmost importance in the theory of the earth. The second class consists of remains, which, by the help of stalactitical concretions, are converted into stone. These are the *exuviæ* of animals, which existed on the very same continents on which we now dwell, and are no doubt the most ancient among their inhabitants, of which any monument is preserved. In comparison of the first class, they must, nevertheless, be considered as of very modern origin.

403. The third class consists of the bones of animals found in the loose earth or soil; these have not acquired a stony character, and their nature

ture appears to be but little changed, except by the progreſs of decompoſition and of mouldering into earth. No decided line can be drawn between the antiquity of this and the preceding claſs, as there may be between the preceding and the firſt. In ſome inſtances, the objects of this third claſs may be coeval with thoſe of the ſecond; in general, they muſt be accounted of later origin, as they are certainly not preſerved in a manner ſo well fitted for long continuance.

404. The animal remains of the ſecond claſs, are generally found in the neighbourhood of limeſtone ſtrata, and are either enveloped or penetrated by calcareous, or ſometimes ferruginous matter. Of this ſort are the bones found in the rock of Gibraltar, and on the coaſt of Dalmatia. The latter are peculiarly marked for their number, and the extent of the country over which they are ſcattered, leaving it doubtful whether they are the work of ſucceſſive ages, or of ſome ſudden cataſtrophe that has aſſembled in one place, and overwhelmed with immediate deſtruction, a vaſt multitude of the inhabitants of the globe. Theſe remains are found in greateſt abundance in the iſlands of Cherſo and Oſero; and always in what the Abbé FORTIS calls an *ocreo-ſtalactitic earth*. The bones are often in the ſtate of mere ſplinters, the broken and confuſed relics of various animals, concreted with fragments of marble

and

and lime, in clefts and chafms of the ftrata*. Sometimes human bones are faid to be found in thefe confufed maffes.

405. A very remarkable collection of bones in this ftate is found in the caves of Bayreuth in Franconia. Many of thefe belong, as is inferred with great certainty from the ftructure of their teeth, to a carnivorous animal of vaft fize, and having very little affinity to any of thofe that are now known. The bones are found in different ftates, fome being without any ftalactitical concretion, and having the calcareous earth ftill united to the phofphoric acid, fo that they belong to the third, rather than the fecond, of the preceding divifions. In others, the phofphoric acid has wholly difappeared, and given place to the carbonic.

The number of thefe bones, accumulated in the fame place, is matter of aftonifhment, when it is confidered, that the animals to which they belonged were carnivorous, fo that more than two can never have lived in the fame cavern at the fame time. The caves of Bayreuth feem to have been the den and the tomb of a whole dynafty of unknown monfters, that iffued from this central fpot to devour the feebler inhabitants of the woods, during a long fucceffion of ages, be-

fore

* Travels into Dalmatia, p. 449.

fore man had ſubdued the earth, and freed it from all domination but his own.

406. The foſſil bones of the ſecond and third claſs, but chiefly of the third, have now afforded matter of conjecture and diſcuſſion for more than a century. The facts with reſpect to them are very numerous and intereſting, but can be conſidered here only very generally.

The remains of this kind, conſiſt of the bones only of large animals, ſo that they have generally been compared with thoſe of the elephant, the rhinoceros, the hippopotamus, or other animals of great ſize. The bones of ſmaller animals have alſo been found, but much more rarely than the other. It is uſually remarked, that the bones thus diſcovered in the earth are larger than thoſe of the ſimilar living animals.

Another general fact concerning theſe remains, is, that they are found in all countries whatſoever, but always in the looſe or travelled earth, and never in the genuine ſtrata. Since the year 1696, when the attention of the curious was called to this ſubject, by the ſkeleton of an elephant dug up in Thuringia, and deſcribed by Tentzelius *, there is hardly a country in Europe which has not afforded inſtances of the

* Phil. Tranſ. vol. xix. p. 757.

the ſame kind. Foſſil bones, particularly grinders and tuſks of elephants, have been found in other places of Germany, in Poland, France, Italy, Britain, Ireland, and even Iceland *. Two countries, however, afford them in greater abundance by far than any other part of the known world; namely, the plains of Siberia in the old continent, and the flat grounds on the banks of the Ohio in the new †.

407. When the bones in Siberia were firſt diſcovered, they were ſuppoſed to belong to an animal that lived under ground, to which they gave the name of the *mammouth*; and the credit beſtowed on this abſurd fiction, is a proof of the ſtrong deſire which all men feel of reconciling extraordinary appearances with the regular courſe of nature. Much ſkill, however, in natural hiſtory was not required to diſcover that many of the bones in queſtion reſembled thoſe of the elephant, particularly the grinders and the tuſks of that animal. Others reſembled the bones of the rhinoceros; and a head of that kind, having the hide

* A grinder of an elephant found in Iceland, is deſcribed by *Bartholinus*, Actor. Hafniens. vol. i. p. 83.

† The foſſil bones on the Ohio are deſcribed in two papers by Mr P. Collinſon, Phil. Tranſ. vol. lvii. p. 464. and 468.

hide preſerved upon it, was found in Siberia, and is ſtill in the imperial cabinet at Peterſburgh.

Pallas has deſcribed the foſſil bones which he found in the muſeum at Peterſburgh, on his being appointed to the ſuperintendence of it, and enumerates, not only bones that belong, in his opinion, to the elephant and rhinoceros, but others that belong to a kind of buffalo, very different from any now known, and of a ſize vaſtly greater *. He has alſo deſcribed, in another very curious memoir, the bones of the ſame kind that he met with in his travels through the north-eaſt parts of Aſia.

The foſſil bones found on the banks of the Ohio, reſemble in many things thoſe of Siberia; like them they are contained in the ſoil or alluvial earth, and never in the ſolid ſtrata; like them too they are no otherwiſe changed from their natural ſtate, than by being ſometimes ſlightly calcined at the ſurface; they are alſo of great ſize, and in great numbers, being probably the remains of ſeveral different ſpecies.

408. Two inquiries concerning theſe bones have excited the curioſity of naturaliſts; firſt, to diſcover among the living tribes at preſent inhabiting

* Novi Comment. Petrop. tom. xiii. (1768) p. 436, and tom. xvii. p. 576, &c.

biting the earth, thoſe to which the foſſil remains may with the greateſt probability be referred; and, ſecondly, to find out the cauſe why theſe remains exiſt in ſuch quantities, in countries where the animals to which they belong, whatever they be, are at preſent unknown. The ſolution of the firſt of theſe queſtions, is much more within our reach than the ſecond, and at any rate muſt be firſt ſought for.

On the authority of ſo eminent a naturaliſt as Pallas, the bones from Siberia may ſafely be referred to the elephant, the rhinoceros, and buffalo, as mentioned above, though perhaps to varieties of them with which we are not now acquainted. With reſpect to the bones of North America, the queſtion is more doubtful, for they have this particular circumſtance attending them, viz. that along with the thigh-bones, tuſks, &c. which might be ſuppoſed to belong to the elephant, grinders are always found of a ſtructure and form entirely different from the grinders of that animal *. Some naturaliſts, particularly M. D'Aubenton, referred theſe grinders to the hippopotamus; but Dr W. Hunter appears to have proved, in a very ſatisfactory manner, that they cannot have

* See Mr Collinſon's papers above referred to, Phil. Tranſ. vol. lvii.

have belonged to either of the animals juſt mentioned, but to a *carnivorous* animal of enormous ſize, the race of which, fortunately for the preſent inhabitants of the earth, ſeems now to be entirely extinct *. The foundation of Dr Hunter's opinion is, that in theſe grinders the enamel is merely an external covering; whereas, in the elephant, and other animals deſtined to live on vegetable food, the enamel is intermixed with the ſubſtance of the tooth †.

409. Though this argument appears to be of conſiderable weight, yet CAMPER, who was greatly ſkilled in comparative anatomy, and who had ſtudied this ſubject with particular attention, was of opinion, that theſe grinders belong to a ſpecies of elephant. This opinion he ſtates in a letter to Pallas, who had found grinders and other bones of this ſame animal, on the weſtern

* Phil. Tranſ. vol. lviii. p. 3, &c.

† A foſſil grinder in the collection of JOHN MACGOWAN, Eſq; of Edinburgh, anſwers nearly to Mr Collinſon's deſcription, and is very well repreſented by the figure which accompanies it. This grinder weighs four pounds one-fourth avoirdupois; the circumference of the *corona* is eighteen inches; the coat of enamel is one fourth of an inch thick; there are five double teeth; in Mr Collinſon's ſpecimen there are only four.

declivity of the Oural mountains *. Camper denies that the animal is carnivorous, becaufe the *incifores*, or canine teeth, are wanting; and he argues farther, from the weight of the head, which may be inferred from the weight of the grinders, that the neck muft have been fhort, and the animal muft have been furnifhed with a *probofcis*. He afterwards abandoned the latter hypothefis, and gave it as his opinion, that the *incognitum* was neither carnivorous, nor a fpecies of the elephant †.

410. Neverthelefs, CUVIER, in a *mémoire* read before the National Inftitute of Paris, maintains, that the foffil bones of the new continent, as well as moft of thofe of the old, belong to certain fpecies of the elephant; of which, at leaft, two do not now exift, and are only known from remains preferved in the ground. He diftinguifhes them thus ‡:

Elephas mammonteus,—maxillâ obtufiore, lamellis molarium tenuibus, rectis.

Elephas Americanus,—molaribus multicufpidibus, lamellis poft detritionem quadri-lobatis.

The latter fpecies, which is meant to include the *animal incognitum*, is faid to have lived, not only

* Acta Acad. Petrop. tom. i. (1777) pars pofterior, p. 213, &c.

† *Ibid.* tom. ii. (1784) p. 262.

‡ Mémoires de l'Inftitut National, Sciences Phyfiques, tom. ii. p. 19., &c.

only in America, but in many parts of the old continent. Yet ſome late inquiries into the ſtructure of the teeth of graminivorous animals, and particularly of the elephant, make it very improbable that the *incognitum* has belonged to this genus *. The grinders of the elephant have been found to conſiſt of three ſubſtances, enamel, bone, and what is called the *cruſta petroſa*, applied in layers, or folds contiguous to one another; and no veſtige of this ſtructure appears in the grinders of the unknown animal of the Ohio †.

 At

* See Mr Home's obſervations on the teeth of graminivorous animals, Phil. Tranſ. 1799. Alſo, An Eſſay on the ſtructure of the teeth, by Dr Blake.

† In a paper inſerted in the fourth volume of the American Philoſophical Tranſactions, an account is given of two different grinders that are found at the Salt-Licks near the Ohio. One of them reſembles the grinder of the elephant, and may have belonged to the elephas Americanus of Cuvier; the other agrees pretty nearly with the grinder of Dr Hunter's *animal incognitum*. The author of the paper thinks that the *animal incognitum* was not wholly carnivorous, as the *inciſores*, or canine teeth, are never found. At the Great Bone-Lick, bones of ſmaller animals, particularly of the buffalo kind, have been diſcovered. The ſaline impregnation of the earth at theſe Licks muſt no doubt have contributed to the preſervation of the bones. Tranſ. American Phil. Soc. vol. iv. (1799) p. 510, &c.

At the ſame time, Dr Hunter's aſſertion, that this animal was carnivorous, is rendered doubtful, not only by the want of *canine* teeth, but alſo from the reſemblance between its grinders and thoſe of the wild boar, which Mr HOME has obſerved to be conſiderable *. The grinder of the boar is ſimilar to that of the elephant, in the extent of the maſticating ſurface, but not at all in the internal ſtructure; and the ſame is true of the tooth of the *animal incognitum*, ſo that a conſiderable probability is eſtabliſhed, that it and the boar are of the ſame genus, and both deſtined to live occaſionally either on animal, or vegetable food.

411. Another *animal incognitum* found in South America has been deſcribed by Cuvier, and appears to be of a different genus from the *incognitum* of the North. Thus, if we include the two *incognita* of America, the *elephas mammonteus*, the unknown buffalo of Pallas, and the great animal of Bayreuth, we have at leaſt five diſtinct genera, or ſpecies of the animal kingdom, which exiſted on our continents formerly, but do not exiſt on them now. The number is probably much greater: Pallas mentions foſſil horns of a gazelle, of an unknown ſpecies; and horns of *deer* are often found, that cannot be referred to any ſpecies now exiſting. Thoſe extinct

* Obſervations on the grinding teeth of the wild boar and *animal incognitum*. Phil. Tranſ. 1801, p. 319.

tinct races have been remarkable for their size: some of the ancient elephants appear to have been three times as large as any of the present *.

412. The inhabitants of the globe, then, like all the other parts of it, are subject to change: It is not only the individual that perishes, but whole *species*, and even perhaps *genera*, are extinguished. It is not unnatural to consider some part of this change as the operation of man. The extension of his power would necessarily subvert the balance that had before been established between the inhabitants of the earth, and the means of their subsistence. Some of the larger and fiercer animals might indeed dispute with him, for a long time, the empire of the globe; and it may have required the arm of a Hercules to subdue the monsters which lurked in the caves of Bayreuth, or roamed on the banks of the Ohio. But these, with others of the same character, were at length exterminated: the more innocent species fled to a distance from man; and being forced to retire into the most inaccessible parts, where their food was scanty, and their migration checked, they may have degenerated from the size and strength of their ancestors, and some species may have been entirely extinguished.

But besides this, a change in the animal kingdom seems to be a part of the order of nature,

 and

* Camper, Nov. Acta Petrop. tom. ii. (1784) p. 257.

and is viſible in inſtances to which human power cannot have extended. If we look to the moſt ancient inhabitants of the globe, of which the remains are preſerved in the ſtrata themſelves, we find in the ſhells and corals of a former world hardly any that reſemble exactly thoſe which exiſt in the preſent. The ſpecies, except in a few inſtances, are the ſame, but ſubject to great varieties. The vegetable impreſſions on ſlate, and other argillaceous ſtones, can ſeldom be exactly recogniſed; and even the inſects included in amber, are different from thoſe of the countries in which the amber is found.

413. Suppoſing, then, the changes which have taken place in the qualities and habits of the animal creation, to be as great as thoſe in their ſtructure and external form, we can have no reaſon to wonder if it ſhould appear, that ſome have formerly dwelt in countries from which the ſimilar races are now entirely baniſhed. The power of living in a different climate, of enduring greater degrees of cold or of heat, or of ſubſiſting on different kinds of food, may very well have accompanied the other changes. Though one ſpecies of elephant may now be confined to the ſouthern parts of Aſia, another may have been able to endure the ſeverer climates of the north; and the ſame may be true of the buffalo or the rhinoceros. In all this no phyſical

ſical impoſſibility is involved; though whether it is a probable ſolution of the difficulty concerning the origin of theſe animal remains, can only be judged of from other circumſtances.

414. If we conſider attentively the facts that reſpect the Siberian foſſil bones, there will appear inſurmountable objections to every theory that ſuppoſes them to be exotic, and to have been brought into their preſent ſituation from a diſtant country.

The extent of the tract through which theſe bones are ſcattered, is a circumſtance truly wonderful. Pallas aſſures us *, that there is not a river of conſiderable ſize in all the north of Aſia, from the Tanais, which runs into the Black Sea, to the Anadyr, which falls into the Gulf of Kamtchatka, in the ſides or bottom of which bones of elephants and other large animals have not been found. This is eſpecially the caſe where the rivers run in plains through gravel, ſand, clay, &c.; among the mountains, the bones are rarely diſcovered. The extent of the tract juſt mentioned exceeds four thouſand miles; and how the bones could be diſtributed over all that extent, by any means but by the animals having

 lived

* De Reliquiis Animalium exoticorum, per Aſiam Borealem repertis.—Nov. Comment. Petrop. tom. xvii, (1772) p. 576.

lived there, it ſeems impoſſible to conceive. No torrent nor inundation could have produced this effect, nor could the bones brought in that way have been laid together ſo as to form complete ſkeletons.

415. One fact recorded by the ſame author, ſeems calculated to remove all uncertainty. It is that of the carcaſe of a rhinoceros, almoſt entire, and covered with the hide, found in the earth in the banks of the river Wilui, which falls into the Lena below Jacutſk *. Some of the muſcles and tendons were actually adhering to the head when Pallas received it. The head, after being dried in an oven, is ſtill preſerved in the muſeum at Peterſburgh. The preſervation of the ſkin and muſcles of this natural mummy, as Pallas calls it, was no doubt brought about by its being buried in earth that was in a ſtate of perpetual congelation; for the place is in the parallel of 64°, where the ground is never thawed but to a very ſmall depth below the ſurface.

But by what means can we account for the carcaſe of a rhinoceros being buried in the earth, on the confines of the polar circle? Shall we aſcribe it to ſome immenſe torrent, which, ſweeping acroſs the deſarts of Tartary, and the mountains of Altai, tranſported the productions of India

* Pallas. *ubi ſupra*, p. 386. Alſo, Voyages de Pallas, tom. iv. p. 131.

dia to the plains of Siberia, and interred in the mud of the Lena the animals that had fed on the banks of the Barampooter or the Ganges? Were all other objections to so extraordinary a supposition removed, the preservation of the hide and muscles of a dead animal, and the adhesion of the parts, while it was dragged for 2000 miles over some of the highest and most rugged mountains in the world, is too absurd to be for a moment admitted. Or shall we suppose that this carcase has been floated in by an inundation of the sea, from some tropical country now swallowed up, and of which the numerous islands of the Indian Archipelago are the remains? The heat of a tropical climate, and the putrescence naturally arising from it, would soon, independently of all other accidents, have stripped the bones of their covering. Indeed this *instantia singularis*, as in every sense it may properly be called, seems calculated for the express purpose of excluding every hypothesis but one from being employed to explain the origin of fossil bones. It not only excludes the two which have just been mentioned, but it excludes also that of Buffon, viz. that these bones are the remains of animals which lived in Siberia, when the arctic regions enjoyed a fine climate, and a temperature like that which southern Asia now possesses. From the preservation of the flesh and hide of this rhinoceros, it is plain, that when the body was buried in the earth,

earth, the climate was much the ſame that it is now, and the cold ſufficient to reſiſt the progreſs of putrefaction.

Pallas takes notice of the inconſiſtency of the ſtate of this ſkeleton, with the hypotheſis of Buffon; but he does not obſerve that the inconſiſtency is equally great between it and his own hypotheſis, the importation of the foſſil bones by an inundation of the ſea, and that fleſh or muſcle muſt have been entirely conſumed long before it could be carried by the waves to the parallel of 64°, from any climate which the rhinoceros at preſent inhabits.

416. The preſence of petrified marine objects in places where ſome of the foſſil bones are found, is no proof that the latter have come from the ſea, though it is produced as ſuch both by Pallas himſelf, and afterwards by Kirwan. Theſe marine bodies are the ſhells and corals that have been parts of calcareous rocks, from which being detached by the ordinary progreſs of diſintegration, they are now contained in the beds of ſand or gravel where the animal remains are buried. They have nothing in common with theſe remains; they are real ſtones, and belong to another, and a far more remote epocha. Such objects being found in the ſame place where the bones lie, argues only that the ſtrata in the higher grounds, from which the gravel has come, are calcareous; and nothing can ſhew in a ſtronger

light

light the neceffity of diftinguifhing the different condition of foffil bodies, united by the mere circumftance of contiguity, before we draw any inference as to their having a common origin. If the marine remains were in the fame condition with the bones; if they were in no refpect mineralized; then the conclufion, that both had been imported by the fea, would have great probability; but without that, their prefent union muft be held as cafual, and can give no infight into the origin of either.

417. On the whole, therefore, no conclufion remains, but that thefe bones have belonged to fpecies of elephants, rhinoceros, &c. which inhabited the very countries where their remains are now buried, and which could endure the feverity of the Siberian climate. The rhinoceros of the Wilui certainly lived on the confines of the Polar circle, and was expofed to the fame cold while alive, by which, when dead, its body has been fo long, and fo curioufly preferved.

Thefe animals may alfo have lived occafionally farther to the fouth, among the valleys between the great ranges of mountains that bound Siberia on that fide. Foffil bones are but rarely found in thefe valleys, probably becaufe they have been wafhed down from thence into the plains. We muft obferve, too, that thofe animals may have migrated with the feafons, and by that means avoided the rigorous winter

winter of the high latitudes. The dominion of man, by rendering ſuch migration to the larger animals difficult or impoſſible, muſt have greatly changed the economy of all thoſe tribes, and narrowed the circle of their enjoyments and exiſtence. The heaps in which the foſſil bones appear to be accumulated in particular places, eſpecially in North America, have a great appearance of being connected with the migrations of animals, and the accidents that might bring multitudes of them into the ſame ſpot.

What holds of Siberia and of North America, is applicable, *a fortiori*, to all the other places where animal remains are found in the ſame condition. Thus we are carried back to a time when many larger ſpecies of animals, now entirely extinct, inhabited the earth, and when varieties of thoſe that are at preſent confined to particular ſituations, were, either by the liberty of migration, or by their natural conſtitution, accommodated to all the diverſities of climate. This period, though beyond the limits of ordinary chronology, is poſterior to the great revolutions on the earth's ſurface, and the lateſt among geological epochas.

NOTE

NOTE XXIII. § 128.

Geology of KIRWAN *and* DE LUC.

418. The two champions of the Neptunian ſyſtem, who have diſtinguiſhed themſelves moſt by their hoſtility to Dr Hutton, are DE LUC and KIRWAN. They have carried on their attack nearly on the ſame plan, and have employed againſt their antagoniſt the weapons both of theology and ſcience. With a ſpirit as injurious to the dignity of religion, as to the freedom of philoſophical inquiry, they have diſregarded a maxim enforced by the authority of Bacon, and by all our experience of the paſt; "*Tanto magis hæc vanitas inhibenda venit et coërcenda, quia, ex divinorum et humanorum male-ſana admixtione, non ſolum educitur philoſophia phantaſtica, ſed etiam religio hæretica. Itaque ſalutare admodum eſt, ſi mente ſobriâ, fidei tantum dentur quæ fidei ſunt* *."

Proceeding

---

* The whole paſſage is deſerving of attention, and it ſeems as if the prophetic ſpirit of Bacon had addreſſed it to the coſmologiſts of the preſent day. "*Peſſima enim res eſt errorum* APOTHEOSIS, *et pro peſte intellectûs habenda eſt, ſi vanis accedat veneratio. Huic autem vanitati nonnulli ex modernis ſummâ levitate ita indulſerunt, ut, in primo capitolo* GENESEOS, *et aliis Scripturis Sacris, philoſophiam naturalem fundari conati ſunt:* Inter VIVA quærentes MORTUA." Nov. Organum, lib. i. aphor. 65.

Proceeding, accordingly, in direct opposition to rules that have never yet been violated with impunity, and mistaking the true object of a theory of the earth, they carry back their inquiries to a period prior to the present series of causes and effects, where, having neither experience nor analogy to direct them, they pretend to be guided by a superior light. They would have us to consider their geological speculations as a commentary on the text of MOSES; they endeavour to explain the action of creative power, and, with indiscreet curiosity, would tear off the veil which the hand of the prophet has so wisely respected. But the veil cannot be torn off, and all that is behind it must be to man as that which never has existed.

419. M. de Luc has nevertheless treated very diffusely of the history of the solar system, previous to the establishment of the present laws of nature, and has dwelt on it with great complacency, and singular minuteness of detail. His tenth letter to LA METHERIE has the following title:

" On the History of the Earth, from the time when that planet was penetrated by *light*, till the appearance of the sun; a portion of time which includes the origin of heat, and of the figure of the earth; of its primeval strata, of the ancient sea, of our continents, as the bottom of that

that ſea, of the great chains of mountains, and of vegetation *."

I muſt confeſs that I am unacquainted with every thing of this letter but the title; and could not eaſily be prevailed on to follow any man who profeſſedly goes out of nature in ſearch of knowledge; who pretends to give the hiſtory of our planetary ſyſtem when there was no ſun, and to enumerate the events which took place between the exiſtence of that luminary, and the exiſtence of light. The abſurdity of ſuch an undertaking admits of no apology; and the ſmile which it might excite, if addreſſed merely to the fancy, gives place to indignation when it aſſumes the air of philoſophic inveſtigation.

420. It ſets, however, in a ſtrong light, the inconſiſtencies that may be obſerved in the intellectual character of the ſame individual, to conſider that the author of this ſtrange and inconſiſtent reverie,

* Journal de Phyſique, tom. 37. (1790) partie 2de, p. 332. As I may not have done juſtice to this extraordinary title, it may be right to preſent it in the original. " Sur l'Hiſtoire de la TERRE, depuis que cette planette fut penetrée de LUMIERE, juſqu'à l'apparition du SOLEIL; eſpace de tems qui renferme les ORIGINES de la *chaleur*, et de la *figure* de notre globe; de ſes *couches primordiales*, de *l'ancienne mer*, de nos *continens*, comme fond de cette mer, de leurs grandes chaînes de *montagnes*, et de la *vegetation*."

reverie is, neverthelefs, an excellent obferver, and well fkilled in experimental inquiries. It will hardly be believed that he who writes the hifto-ry of the earth before the formation of the fun, is verfed in the principles of inductive reafoning; and that he has added much to the ftock of geo-logical knowledge, having obferved accurately, and defcribed with great perfpicuity and can-dour. His *Lettres Phyfiques* are full of valuable and juft obfervations, though accompanied with reafonings that do not feem always entitled to the fame praife; and in another work he has fuc-ceeded where many men of genius had failed, and has made confiderable improvements in a branch of the mathematics, without borrowing almoft any affiftance from the principles of that fcience *.

421. Some of the fame obfervations apply to Mr Kirwan. His Geological Effays have alfo for their object to explain the firft origin of things; and to fay that he has not fucceeded, in an at-tempt where no man ever can fucceed, im-plies no reproach on the execution of his work, whatever it may do on the defign. We have indeed no criterion by which the execution of it can be eftimated: what would in any other place be a blemifh, may be here deferving of praife; and if the work is full of confufion and perplexity

* Effai fur les Modifications de l'Atmofphere.

perplexity, these are qualities inherent in the subject which it is intended to describe. It were, no doubt, to be wished, that after emerging into the regions of day, Mr Kirwan had been as successful in copying the beauty and simplicity of nature, as in representing the disorder and inconsistency of the chaotic mass. But his cosmology is without unity in its principles, or consistency in its parts: the causes introduced, are, for the most part, such as will account for one set of appearances just as well as for another; or, if any of them is likely to prove inadequate to the effect ascribed to it, a new and arbitrary hypothesis is always ready to come to its assistance. The information given is seldom exact: a multitude of facts brought together, without the order and discussion essential to precise knowledge; and an infinity of quotations, amassed without criticism or comparison, afford proofs of extensive reading, but of the most hasty and superficial inquiry. Thus we have seen passages from Ulloa and Frisi, produced in support of opinions, which, when fairly stated, they had the most direct tendency to overthrow.

422. In one respect, the geological writings of Kirwan are far inferior to De Luc's: They are evidently the productions of a man who has not seen nature with his own eyes; who has studied

mineralogy in cabinets, or in books only; but who has ſeldom beheld foſſils in their native place. With the balance in his hand, and the external characters of WERNER in his view, he has examined minerals with diligence, and has diſcovered many of thoſe marks which ſerve to aſcertain their places, in a ſyſtem of artificial arrangement. But to *reaſon* and to *arrange* are very different occupations of the mind; and a man may deſerve praiſe as a mineralogiſt, who is but ill qualified for the reſearches of geology.

423. The ſame hurry and impatience are viſible in the manner in which his argument againſt Dr Hutton is uſually conducted. He has ſeldom been careful to make himſelf maſter of the opinions of his adverſaries; and what he gives as ſuch, and directs his reaſonings againſt, have often no reſemblance to them whatſoever. Without any intention to deceive others, but deceived himſelf, he uſually begins with miſrepreſenting Dr Hutton's notions, and then proceeds to the refutation of them. In this imaginary conteſt, it will readily be ſuppoſed, that he is in general ſucceſsful: when a man has the framing both of his own argument, and that of his antagoniſt, he muſt be a very unſkilful logician if he does not come off with the advantage.

424. It

424. It is but juſtice, however, to the Neptuniſts, to acknowledge, that they are not all liable to the cenſure of beginning their reſearches from a period antecedent to the exiſtence of the laws of nature. This abſurdity does not, ſo far as I know, infect the ſyſtem of Werner. That mineralogiſt has not propoſed to explain the firſt origin of things, though he has ſuppoſed, at ſome former period, a condition of the globe very unlike the preſent, viz. the entire ſubmerſion of the ſolid under the fluid part.

Note xxiv. § 129.

Syſtem of Buffon.

425. The affinity of Dr Hutton's theory to that of Buffon, is nothing more than what ariſes from their making uſe of the ſame agents, viz. fire and water, in producing the preſent condition of the earth's ſurface. In almoſt all other reſpects the two theories are extremely different. The order in which tho e agents are employed in them, is directly oppoſite, as has already been remarked; Buffon introducing the action of fire firſt, and of water only in the ſecond place, to waſte and deſtroy mineral bodies,

and afterwards to difpofe them anew, and arrange them into ftrata. He makes no provifion for the confolidation of thefe ftrata, nor any for their angular elevation; he has no means of explaining the unftratified rocks; nor any, but one extremely imperfect, for explaining the inequalities of the earth's furface.

Again, Buffon miftook, in fome degree, the true object of a theory of the earth; and though he did not go back, like the geologifts juft named, to a time when the laws of nature were not fully eftablifhed, he begins from a condition of things too unlike the prefent to be the bafis of any rational fpeculation. He does not, indeed, undertake to examine the ftate of our planetary fyftem before the fun exifted; for from fuch extravagance, even when moft difpofed to indulge his fancy, he would furely have revolted. But he treats of the world, when the earth and the planets had juft ceafed to be a part of the fun, and were newly detached from the body of that luminary *.

This hypothefis concerning the origin of the planets, contrived chiefly to account for the circumftance

* According to Buffon, the granite is the true folar matter, unchanged but by its congelation.

cumſtance of their motion being all in the ſame direction, and in other reſpects not only unſupported, but even inconſiſtent with the principle of gravitation, has nothing in common with a theory, confined as Dr Hutton's is, within the field which muſt for ever bound our inquiries, and not venturing to ſpeculate about the earth, when in a condition totally different from the preſent.

426. In what relates to the future, the two ſyſtems are not more like than in what relates to the paſt. Buffon repreſents the cooling of our planet, and its loſs of heat, as a proceſs continually advancing, and which has no limit, but the final extinction of life and motion over all the ſurface, and through all the interior, of the earth. The death of nature herſelf is the diſtant but gloomy object that terminates our view, and reminds us of the wild fictions of the Scandinavian mythology, according to which, *annihilation* is at laſt to extend its empire even to the gods. This diſmal and unphiloſophic viſion was unworthy of the genius of Buffon, and wonderfully ill ſuited to the elegance and extent of his underſtanding. It forms a complete contraſt to the theory of Dr Hutton, where nothing is to be ſeen beyond the continuation of the preſent order; where no latent ſeed of evil threatens final deſtruction to the whole; and where the

movements are ſo perfect, that they can never terminate of themſelves. This is ſurely a view of the world more ſuited to the dignity of NATURE, and the wiſdom of its AUTHOR, than has yet been offered by any other ſyſtem of coſmology.

427. I have often quoted Buffon in the courſe theſe *Illuſtrations*, and moſt commonly for the purpoſe of combating his opinions; but I am very ſenſible, nevertheleſs, of the obligations under which he has laid all the ſciences connected with the natural hiſtory of the earth.

The extent and variety of his knowledge, the juſtneſs of his reaſonings, the greatneſs of his views, his correct taſte, and manly eloquence, qualified him, better, perhaps, than any other individual, to compoſe the Hiſtory of Nature. The errors into which he has fallen, are almoſt all the unavoidable conſequences of the circumſtances in which he was placed; and if their amount is eſtimated by the proportion that they bear to the general excellence of the work, they will be reckoned but of ſmall account. Buffon began to write when many parts of natural hiſtory had made but little progreſs; when the quantity of authentic information was ſmall, and when ſcientific and correct deſcription was hardly to be found. Many of the greateſt and moſt important facts in geology were quite unknown, and

and ſcarcely any part of the mineral kingdom had been accurately ſurveyed; and, with ſuch materials as this ſtate of things afforded, it is not wonderful if ſome parts of the edifice he erected have not proved ſo ſolid and durable as the reſt. Had he appeared ſomewhat later; had he been farther removed from the time when reaſonings *a priori* uſurped the place of induction; and had he been as willing to correct the errors into which he had been betrayed by imperfect information, as he was ingenious in defending them, his work would probably have reached as great perfection, as it is given for any thing without the ſphere of the accurate ſciences to attain. If he had examined the natural hiſtory of the earth more with his own eyes, and been as careful to delineate it with fidelity as force; if he had liſtened with greater care to the philoſophers around him; had he attended to the demonſtrations of NEWTON more, and deſpiſed the arrangements of LINNÆUS leſs; he would have produced a work, as ſingular for its truth as for its beauty, and would have gone near to merit the eulogy pronounced by the enthuſiaſm of his countrymen, MAJESTATI NATURÆ PAR INGENIUM.

NOTE XXV. § 130.

Figure of the Earth.

428. That the earth is a ſpheroidal body, compreſſed at the poles, or elevated at the equator, is a fact eſtabliſhed by many accurate experiments; and though theſe experiments do not exactly coincide, as to the degree of oblateneſs which they give to that ſpheroid, they agree ſufficiently to put it beyond all diſpute, that the earth, though ſolid, has nearly the ſame figure which it would aſſume if fluid, in conſequence of its rotation on its axis.

Now, it is not at all obvious, to what phyſical cauſe this phenomenon is to be aſcribed. The earth, as it exiſts at preſent, has none of the conditions that render the aſſumption of the figure of equilibrium in any way neceſſary to it. Conſtituted as it is, its parts cohere with forces incomparably too great to obey the laws of ſtatical preſſure, or to aſſume any one figure rather than another, on account of the centrifugal tendency which reſults from its revolution on its axis. There is no neceſſity that its ſuperficies ſhould be every where level, or perpendicular to the direction of gravity, nor that every two columns,

lumns, ſtanding on the ſame baſe, any where within it, and reaching from thence to any two points of the ſurface, ſhould be of ſuch weights as preciſely to balance one another. Neither of theſe, indeed, is at all conformable to fact. They are, however, the very ſuppoſitions on which the determination of the ſpheroid of equilibrium is founded; and as they certainly do in no degree belong to the earth, it ſeems ſtrange that the reſult deduced from them ſhould be in any way applicable to it. This coincidence remains, therefore, to be explained; and it muſt greatly enhance the merit of any geological ſyſtem, if it can connect this great and enigmatical phenomenon with the other facts in the natural hiſtory of the earth.

429. To eſtabliſh ſuch a connection, has, accordingly, been a favourite object with geologiſts, whether they have embraced the Neptunian or Vulcanic theory: both have thought that they were entitled to ſuppoſe the primeval fluidity of the globe, the one by water, and the other by fire; and in whatſoever way that fluidity was produced, the reſult of it could be no other than the ſpheroidal figure of the whole maſs, agreeably to the laws of hydroſtatics. If in this fluid ſtate the earth was homogeneous, the ſpheroid would be accurately elliptical, and the compreſſion at the poles would be $\frac{1}{230}$ of the radius of the

the equator ; if the fluid was denſer toward the centre, the flattening would be leſs : and in either caſe, the body, as it acquired ſolidity, may be ſuppoſed to have retained its ſpheroidal figure with little variation. But though the fluidity of the earth will account for the phenomenon of its oblate figure, it may reaſonably be queſtioned, whether this fluidity can be admitted, in conſiſtency with other appearances. According to what is eſtabliſhed above, none of the appearances in the mineral kingdom indicate more than a partial fluidity in any former condition of the earth. The preſent ſtrata, made up as they are of the ruins of former ſtrata, though ſoftened by heat, have not been rendered fluid by it, and have even poſſeſſed their ſoftneſs in parts, and in ſucceſſion, not altogether, nor at the ſame time.

The unſtratified, and more cryſtallized ſubſtances, were caſt in the boſom of others, which were ſolid at the time when they were fluid. In all this, therefore, there is no indication of a fluidity prevailing through the whole maſs, or even over the whole ſurface of the earth, and therefore nothing that can explain the ſpheroidal figure which it has acquired. The ſuppoſition, then, of the entire body of the earth, or even of its external cruſt, having been fluid, though it might account for the compreſſion at the

the poles, does not connect that fact with the other facts in the natural history of the globe, and fails, therefore, in the point most essential to a theory. It is liable, also, to other objections: whether it be conceived to have proceeded from fire or from water; whether it has happened on the principles of Buffon or of Werner.

430. First, let us suppose that the fluidity of the earth, or of the external crust of it, at least to a certain depth, proceeded from a solution of the whole in the waters of the ocean; and, waving all the objections that have been stated to this hypothesis, on account of the absolute insolubility of many mineral substances in water, let us suppose them all soluble in a certain degree, and let us compute the quantity of the menstruum, which, on the suppositions most favourable to the system, must have been required to this great geologico-chemical operation.

The siliceous earth, though not soluble in water *per se*, yet, after being dissolved in that fluid by means of an alkali, was found by Dr Black, in his analysis of the Geyser water, to remain suspended in a quantity of water, between 500 and 1000 times its own weight. This is one of the facts most favourable to the Neptunian theory; and that every advantage may be given to that theory, we shall take the least of the numbers just mentioned, and suppose that siliceous earth

earth may be diſſolved or ſuſpended in 500 times its weight of water.

Taking this for the extreme degree of inſolubility of mineral ſubſtances, (though there are many of which the inſolubility is abſolute, or, to ſpeak in the language of calculation, infinitely great), we may ſuppoſe the inſolubility of all the reſt, or the quantities of water in which they are diſſolved, to be ranged in a deſcending ſcale from 500 to 0, the extreme degree of deliqueſcence. Then, taking the arithmetical mean between theſe extremes, it will give us 250, as the proportion of water in which mineral ſubſtances may at an average be diſſolved. But this average is much leſs than the truth ; for the quantity of ſiliceous earth is great in compariſon of any of the reſt, and the mineral ſubſtances that are extremely ſoluble in water are but in a ſmall quantity ; therefore, when we ſuppoſe mineral bodies, at a medium, to be ſoluble in 250 times their own weight of water, we make a ſuppoſition extremely favourable to the Neptunian ſyſtem.

431. This is the proportion between the *weight* of the ſolvent, and of the ſubſtances held in ſolution : to have the proportion of their *bulks*, we may ſuppoſe the ſpecific gravity of mineral bodies in general to be to that of water as 5 to 2, and then we have the ratio of bulks, that of

250

250×5 to 2×1, or of 625 to 1. It follows, then, that minerals in general cannot be ſuppoſed ſoluble in leſs than 625 times their bulk of water.

432. Again, it muſt be allowed to the Neptuniſts, that the fluidity of the whole earth is not neceſſary to account for its aſſuming the ſpheroidal figure. It is ſufficient if the whole of that cruſt or ſhell of matter was fluid, which is contained between the actual ſurface of the terreſtrial ſpheroid, and the ſurface of the ſphere inſcribed within it; that is, of the ſphere which has for its diameter the polar axis of the earth. The whole of the minerals which compoſe this ſhell, muſt at leaſt have been diſſolved in water, and have formed the chaotic maſs of Mr Kirwan. The volume of the water required for this was not leſs than 625 times the bulk of the ſpheroidal ſhell that has juſt been mentioned.

But, aſſuming the difference between the polar axis and the equatorial diameter to be $\frac{1}{300}$ of the latter, which is the ſuppoſition moſt agreeable to the phenomena, it is eaſy to ſhew that the magnitude of the above ſpheroidal ſhell, or the difference between the ſolid content of the earth, and the ſphere inſcribed in it, is greater than $\frac{1}{151}$, and leſs than $\frac{1}{150}$ of the whole earth; ſo that

that the earth is leſs than 151 times the ſpheroidal ſhell.

The volume of the water, therefore, neceſſary to hold in ſolution the materials of this ſhell, is to the volume of the whole earth as 625 to 151, or in a greater ratio than that of four to one: and ſuch, therefore, at the very leaſt, is the quantity of water which Mr Kirwan ſuppoſes, after it ceaſed to act in its chemical capacity, to have retired into caverns in the interior of the earth. Thus the Neptuniſts, in their account of the ſpheroidal figure of the earth, are reduced to a cruel dilemma, and are forced to chooſe between a phyſical and a mathematical impoſſibility.

If we would inquire whether the opinion of the igneous origin of minerals, as commonly received by the Vulcaniſts, is capable of affording a better ſolution of this difficulty, the theory of M. de Buffon is the firſt that preſents itſelf.

433. That philoſopher conſiders the exiſtence of the ſpheroidal figure as a proof that the whole of the earth muſt have been originally fluid; and as the fluidity of the whole can only be aſcribed to fuſion, he has ſuppoſed that the earth was originally a maſs of melted matter ſtruck off from the ſun by the colliſion of a comet; and that this maſs, when made to revolve on its axis.

axis, put on a ſpheroidal figure, which it has retained, though now cooled down to congelation.

This ſyſtem need not be conſidered in detail; the foundation of it is laid in ſuch defiance of the principles of geometry and mechanics, that the architect, notwithſtanding all the fertility of his invention, and all the reſources of his genius, was never able to give any ſolidity to the ſtructure.

But it will be ſaid, that we may take a part of the ſyſtem, without venturing on the whole, and may ſuppoſe that the earth, or at leaſt the external cruſt of it, has been fluid by fire, though we do not inquire into the cauſe of this fire, or into the manner in which it was produced.

It is indeed true, that, when this is done, we have not the ſame ſort of abſurdity to encounter that we met with in the Neptunian ſyſtem, and that the Vulcanic theory does not, like it, come into direct colliſion with an axiom of geometry. There are, nevertheleſs, great objections to it; for though all the phenomena of the mineral kingdom atteſt a fluidity of igneous origin, yet it is a fluidity that was never more than partial; and though it has been over all the earth, has been over it in ſucceſſion only. Beſides, we are not entitled

to

to aſſume the exiſtence, and again the diſappearance of ſuch a great quantity of heat, without aſſigning ſome cauſe for the change.

434. Since, then, neither the hypotheſis of the Neptuniſts or the Vulcaniſts, affords any good explanation of the figure of the earth, or ſuch a one as can connect it with the other appearances in its natural hiſtory, it remains to inquire, whether the ſyſtem that ſuppoſes a partial and ſucceſſive fluidity, like Dr Hutton's, has any reſource for explaining this great phenomenon.

Of this ſubject Dr Hutton has not treated; and when I was firſt made acquainted with his ſyſtem, it appeared to me a very ſerious objection to it, that it did not profeſs to give an explanation of ſo important a fact as the oblate figure of the earth: On conſidering the matter more cloſely, however, I found that there were principles contained in it from which a very ſatisfactory ſolution (and, I think, the only ſatisfactory ſolution) of that difficulty might be deduced. This ſolution I ſhall endeavour to explain, in as far, at leaſt, as is neceſſary for the purpoſe of general illuſtration.

It is laid down in Dr Hutton's theory, that the ſurface of the earth is perpetually changed by the *detritus* of the land; and that from the materials

materials thus afforded, new horizontal ſtrata are perpetually formed at the bottom of the ſea. If this be true, and if the alternations of decay and renovation have been often repeated, it is certain, that the figure of the earth, whatever it may have originally been, muſt be brought at length to coincide with the ſpheroid of equilibrium.

435. Here it is neceſſary to remark, that the expreſſions, *figure of the earth*, and *ſurface of the earth*, are each of them occaſionally taken in two different ſenſes.

The ſurface of the earth, in its moſt obvious ſenſe, is that which bounds the whole earth, and includes all its inequalities ; it is a ſurface extremely irregular, riſing to the tops of the mountains, deſcending to the bottoms of the valleys, and having the continuity of its curvature often interrupted, or ſuddenly changed. This may be called the *actual* ſurface, and the figure bounded by it, the *actual* figure, of the earth.

The ſurface of the earth, in another ſenſe, is one that is every where horizontal, and is the ſame which water aſſumes when at reſt.

This ſuperficies is determined by the circumſtance of its being conſtantly perpendicular to the direction of gravity ; it is the ſurface marked out by levelling, and may be ſuppoſed to be continued from the ſea, through the

 interior

interior of the land, till it meet the ſea again. The figure bounded by this horizontal ſurface, may properly be called the *ſtatical* figure of the earth.

When it is ſaid that the figure of the earth is an oblate ſpheroid, it is the ſtatical, not the actual figure which is meant; and the degrees of the meridian which aſtronomers meaſure, are alſo referred to the ſuperficies of the former.

436. Suppoſe now a body like the earth, but with its actual figure infinitely more irregular, having a ſea circumfuſed around it, the water will deſcend into the loweſt ſituations, and will ſo arrange itſelf, that its ſurface ſhall be perpendicular every where to the plumb-line, or to the direction of gravity, in which ſtate only it can remain at reſt. The figure of the ſuperficies which the ſea muſt thus take will be of a continuous curvature, and will return into itſelf; though it may, if the actual figure is very irregular, be far either from a ſphere or a ſpheroid. If, however, we ſuppoſe the ſolid parts of this maſs ſubject to be diſſolved or worn away, and carried down to the ocean, there will be a tendency to give to the whole body the ſame figure that it would have aſſumed, if it had been entirely fluid, and ſubject to the

laws

laws of hydroſtatics. This tendency is the reſult of two principles.

437. Let us ſuppoſe the body juſt deſcribed to have no rotation, ſo that the particles of it are actuated only by the forces of coheſion and of attraction.

It is then clear, that every particle taken away by attrition from the parts above the level of the ſea, and depoſited under the ſurface of it, makes the general figure more compact, bringing the remoter parts nearer to the centre of gravity of the whole; ſo that, in time, if the body is homogeneous, all the points of the ſurface will become equally diſtant from that centre. Thus the *actual* figure changes continually, and approaches nearer to the *ſtatical.*

While this change is going forward in the actual figure, there is another produced on the ſtatical, that tends very much to accelerate the final coincidence of the two.

The effect of the inequalities of the land, that riſe above the horizontal ſurface, is, by their attraction, to render the parts of that ſurface immediately under them, more convex, *cæteris paribus*, than the reſt. Again, where there are parts of extraordinary depth in the ſea, that is, where the ſolid and denſer parts are far removed from the ſurface of the ocean, the curvature of the ſuperficies of the ſea is thereby diminiſhed, and that ſuperficies is rendered leſs convex than it would be if the ſea were ſhallower. Theſe propoſitions are both capable of ſtrict mathematical demonſtration. Hence the taking away of any particle of matter from the top of a mountain tends to diminiſh the curvature of the horizontal ſurface under the mountain, where it is greateſt; and the depoſition of the ſame particle at the bottom of the ſea, tends to increaſe the curvature of this ſuperficies where it is leaſt. The general tendency, therefore, being to increaſe the curvature where it is leaſt, and to diminiſh it where it is greateſt, muſt be to bring about an uniform curvature throughout, that is, a ſpherical figure. Thus, by the waſte and ſubſequent ſtratification of the land, the direction of gravity is continually altered; it is more and more concentrated, and the figure brought nearer to that which a fluid would aſſume.

438. If now we ſuppoſe the body to revolve on its axis, all other things remaining as before, the ſurface bounding the ſea will become different from what it was in the former caſe, and will be more ſwelled out toward the middle or equatorial regions. The land above the level of the ſea will ſtill, as before, be worn down and depoſited in the bottom of the ſea, ſo as to form ſtrata nearly parallel to its ſurface: the tendency, therefore,

is

is to render the real figure of the planet nearer to the ſtatical. At the ſame time the *ſtatical* figure is changed, as explained above; ſo that the two figures mutually approach, and the limit, or ultimate figure to which they tend, is one over which the ocean might be diffuſed every where to the ſame depth, for then the cauſes of change would entirely ceaſe. But this figure is no other than the ſpheroid of equilibrium, which, therefore, is the effect which the waſte and reconſolidation of the land would neceſſarily produce, if the proceſs were continued indefinitely, without interruption. In this, as in many other inſtances, when a body is ſubject to the action of cauſes by which its form is *gradually* changed, the figure beſt adapted to reſiſt thoſe changes, is the figure which the changes themſelves ultimately produce.

Alſo, whatever be the irregularities of denſity, the tendency to a change of figure will not ceaſe till the body is moulded into that particular ſpheroid which admits of being covered with water every where to the ſame depth *. Thus it

* In the ſame manner as a tranſition is thus made from an irregular figure to a ſpheroid of equilibrium, ſo, if the actual figure were at firſt more ſimple than the ſpheroid, it would ſtill be changed into this laſt by degrees.

Let

it appears, that a ſolid of an irregular figure, and of irregular denſity, provided it be in part covered

Let us conceive, for inſtance, that the earth is at reſt, and is a perfect ſphere of ſolid matter, ſurrounded by an ocean every where of equal depth, for example, of one mile. Then, if a rotatory motion be communicated to it, ſo that it ſhall revolve on its axis in twenty-four hours, in conſequence of the centrifugal force, the water circumfuſed about the ſphere will immediately riſe up under the equator, and will become part of a ſpheroidal ſurface, (not elliptical, but nearly ſo), the equatorial diameter of which is greater than the polar axis, in the ratio of 588 to 577. By this means the water will be accumulated at the equator to the depth of nearly 2.5 miles, and form a zone ſurrounding the earth, and extending about 37° on each ſide of the equator. The remainder of the ſurface will be left dry, forming two vaſt circumpolar continents, that reach 53° on every ſide of the poles, and that are elevated in the middle more than four miles above the level of the ſea.

Such would be the ſtate of our globe, on the hypotheſis above laid down; and, if there were no waſte or deſtruction of the land, this order of things would be permanent, and neither the ſolid nor fluid part of the maſs could ever acquire any other figure than that which has been deſcribed. But, if the ſame laws be ſuppoſed to regulate the action of the atmoſphere in thoſe circumſtances, that do actually regulate it according to the preſent conſtitution of the globe, the vapours raiſed up from the ſurface of the ſea, would be carried by the winds over

vered with water; and be at the ſame time ſubject to waſte above the ſurface of the ſea, and reconſolidation under it, has a tendency to acquire, in time, the ſame figure that it would have acquired had it been entirely fluid.

over the land, where they would be condenſed and precipitated in rain. Thus, all the agents of deſtruction would be let looſe on the two great circumpolar continents; rivers would be formed; the land would become deeply interſected by ravines; thoſe ravines would gradually open into wide valleys; the maſſes of greateſt reſiſtance would be ſhaped into hills and mountains: and from a ſuperficies originally ſmooth and uniform, the ſame inequalities would be produced which at preſent diverſify the ſurface of the earth.

While the parts of the ſphere without the ſpheroid are thus continually diminiſhed, the looſe earth and ſand waſhed down from them, will be depoſited at the bottom of the ſea, and will form ſtrata parallel to the ſurface of the ſuperincumbent water. The actual and ſtatical figure are thus brought nearer one another; and, at the ſame time the ſtatical is changed, on the principle already explained (the change in the direction of gravity), and is made continually to approximate to a ſtate, which when it has attained, no farther change can take place, viz. an oblate elliptic ſpheroid, of which the ſurface is perpendicular to the direction of gravity, having the equatorial diameter to the polar axis in the ratio of 230 to 229.

439. In the preceding reaſonings, we have ſuppoſed the proceſs of decay and ſubſequent ſtratification to be carried on without interruption, till the whole of the land is covered by the ſea. This ſuppoſition is uſeful for explaining the nature of the forces which have determined the figure of the earth; but there is no reaſon to think that it has ever been realized in its full extent, the elevation of ſtrata from the bottom of the ſea interrupting the progreſs, and producing new land in one place as the old decays in another. The very ſame land alſo, which is waſted at its ſurface, may perhaps be lifted up by the forces that are placed under it; or it may be let down, undergoing alterations of its level, from cauſes that we do not perceive, but of which the action is undoubted (§ 387). But notwithſtanding theſe interruptions, the general tendency to produce in the earth a ſpheroidal figure may remain, and more may be done by every revolution, to bring about the attainment of that figure than to cauſe a deviation from it. This figure, therefore, though never likely to be perfectly acquired, will be the *limiting* or *aſymptotic* figure, if it may be ſo called, to which the earth will continually approach.

440. If the preceding concluſions are juſt, and if the figure of equilibrium is only an aſymptotic figure, to which that of the earth may approximate,

proximate, but cannot perfectly attain, we are not to be surprised if considerable deviations from it are actually observed. This has accordingly happened, insomuch, that the results deduced from the most accurate measurement of degrees of the meridian, differ from one another, in the oblateness they give to the earth, by nearly one-half of the quantity to be determined. When we compare the degrees measured in France, and in some other countries of Europe, with those measured in Peru, we obtain for the compression at the poles, less than $\frac{1}{300}$ of the radius of the earth. But when we compare the degrees measured in France with one another, and with those lately measured in England, we find that they are best represented by a spheroid that has its compression $\frac{1}{150}$ of its semi-axis *. There is reason to think, therefore, that the meridians are not elliptical; and other observations seem to show, that they are not even similar to one another; or that the earth is not, strictly speaking, a solid of revolution; so, also, the comparison of the degree measured at the Cape of Good Hope, with those measured on the opposite

* Exposition du Système du Monde, par La Place, p. 61. 2d edit.

oppoſite ſide of the equator, creates a ſuſpicion, that the northern and ſouthern hemiſpheres are not perfectly alike, and that the earth is not equally compreſſed at the Arctic and the Antarctic poles. Theſe irregularities, though they do not affect the general fact of the earth's compreſſion at the poles, ſhew that the true ſtatical figure is but imperfectly attained; and though this may be accounted for, without having recourſe to the principles involved in our theory, it is in a manner very unſatisfactory, and, by help of ſuppoſitions, not at all conſiſtent with the original fluidity aſcribed to the whole maſs, or to the exterior cruſt of the earth.

441. As the principles here laid down explain how a ſolid body may attain very nearly the figure which a fluid would acquire in order to preſerve its parts in equilibrio; and ſince the oblate figure belongs to other of the planets as well as the earth, and the globular to all the great bodies of the univerſe, this ſuggeſts an analogy that goes deep into the economy of nature, and extends far beyond the limits within which the mineralogiſt is wont to confine his ſpeculations.

442. That no very irregular figure is found among the planetary bodies, may therefore be conſidered as a proof of the univerſality of that ſyſtem of waſte and reconſolidation that we have been

been endeavouring to trace in the natural hiſtory of the earth. A farther proof of the ſame ariſes from conſidering, that for every given maſs of matter, having a given period of rotation, there are two different ſpheroids that anſwer the conditions of eſtabliſhing an equilibrium among its parts, the one near to the ſphere, and the other very diſtant from it, and ſo oblate as to have a lenticular form. Thus the earth, ſuppoſing it homogeneous, might either be in equilibrio, by means of the figure which it actually has, or of one in which the polar was to the equatorial diameter as 1 to 768. The ſame is true of the other planets; and yet we no where find that this highly compreſſed ſpheroid is actually employed by nature. The reaſon, no doubt, is, that in ſo oblate a ſpheroid, the equilibrium between the gravitating and the centrifugal force is of the kind that does not re-eſtabliſh itſelf when diſturbed; ſo that the parts let looſe, and not kept in their place by firm coheſion, would fly off altogether. In ſuch a body, the waſte at the ſurface would lead to an entire change of form, and therefore the conſtitution here ſuppoſed could not be permanent.

443. In the ſyſtem of Saturn, we have a great deviation from the general order, which, nevertheleſs, has led to a very unexpected verification of ſome of the concluſions deduced above. A principle

principle extremely like that which is the basis of all the foregoing reasonings, led one of the greatest philosophers of the present age to discover the revolution of Saturn's ring on its axis, and even to determine the velocity of that revolution, such as it has been since found by observation. LA PLACE, laying it down as a maxim, that nothing in nature can exist, where there are causes of change, not balanced or compensated by other causes *, concluded, that the parts of the ring must be held from falling down to the body of the planet by some other force than their mere cohesion to one another. Were it otherwise, every particle detached from the ring, by any means, must descend in a straight line, almost perpendicular to the surface of Saturn; and the final destruction of the ring must be inevitable. The only force that could balance this effect of gravitation, seemed to be a centrifugal force, arising from the rotation of the ring on an axis passing through its centre, and perpendicular to its plane. La Place proceeded to inquire what celerity of rotation was adequate to this effect, and found that one of ten hours and a quarter would be required, which is almost precisely the time afterwards determined by Dr HERSCHEL from actual observation. If, with this rotation,

* La Place, *ubi supra*, p. 242.

rotation, the ring is a ſolid annulus generated by the rotation of a very flat ellipſis about a given point in its greater axis, coinciding with the centre of Saturn, it may be ſo conſtituted, that the attraction of Saturn, combined with the centrifugal force, may produce a force perpendicular to its ſurface, and may enable detached parts to remain at reſt, animals, for inſtance, to walk on its ſurface, and fluids to be *in equilibrio*. The ſyſtem of Saturn is thus fortified againſt the lapſe of time, as effectually as that of the earth itſelf; and the means by which this is accompliſhed, ſeem to prove, that the weapons which time employs, are in both caſes the ſame, viz. the ſlow wearing and decompoſition of the ſolid parts. This ſlow wearing may have produced the figure by which its action is moſt effectually reſiſted.

444. Thus Dr Hutton's theory of the earth comes at laſt to connect itſelf with the reſearches of phyſical aſtronomy. The concluſion to be drawn from this coincidence is to the credit of both ſciences. When two travellers, who ſet out from points ſo diſtant as the mineralogiſt and the aſtronomer, and who follow routes ſo different, meet at the end of their journey, and agree in their report of the countries through which they have paſſed, it affords no ſlight preſumption, that they have kept the right way,

and

and that they relate what they have actually ſeen.

Note XXVI. § 133.

Prejudices relating to the Theory of the Earth.

445. Among the prejudices which a new theory of the earth has to overcome, is an opinion, held, or affected to be held, by many, that geological ſcience is not yet ripe for ſuch elevated and difficult ſpeculations. They would, therefore, get rid of theſe ſpeculations, *by moving the previous queſtion*, and declaring that at preſent we ought to have no theory at all. We are not yet, they allege, ſufficiently acquainted with the phenomena of geology; the ſubject is ſo various and extenſive, that our knowledge of it muſt for a long time, perhaps for ever, remain extremely imperfect. And hence it is, that the theories hitherto propoſed have ſucceeded one another with ſo great rapidity, hardly any of them having been able to laſt longer than the diſcovery of a new fact, or a fact unknown when it was invented. It has proved inſufficient to connect this fact with the phenomena already known, and has therefore been juſtly abandoned. In this manner, they ſay, have paſſed away the theories of Woodward, Burnet, Whiſton, and even of Buffon;

fon; and fo will pafs, in their turn, thofe of Hutton and Werner.

446. This unfavourable view of geology, ought not, however, to be received without examination; in fcience, prefumption is lefs hurtful than defpair, and inactivity is more dangerous than error.

One reafon of the rapid fucceffion of geological theories, is the miftake that has been made as to their object, and the folly of attempting to explain by them the firft origin of things. This miftake has led to fanciful fpeculations that had nothing but their novelty to recommend them, and which, when that charm had ceafed, were rejected as mere fuppofitions, incapable of proof. But if it is once fettled, that a theory of the earth ought to have no other aim but to difcover the laws that regulate the changes on the furface, or in the interior of the globe, the fubject is brought within the fphere either of obfervation or analogy; and there is no reafon to fuppofe, that man, who has numbered the ftars, and meafured their forces, fhall ultimately prove unequal to this inveftigation.

447. Again, theories that have a rational object, though they be falfe or imperfect in their principles, are for the moft part approximations to the truth, fuited to the information at the time

time when they were propofed. They are fteps, therefore, in the advancement of knowledge, and are terms of a feries that muft end when the real laws of nature are difcovered. It is, on this account, rafh to conclude, that in the revolutions of fcience, what has happened muft continue to happen, and becaufe fyftems have changed rapidly in time paft, that they muft neceffarily do fo in time to come.

He who would have reafoned fo, and who had feen the ancient phyfical fyftems, at firft all rivals to one another, and then fwallowed up by the Ariftotelian; the Ariftotelian phyfics giving way to thofe of Des Cartes; and the phyfics of Des Cartes to thofe of Newton; would have predicted that thefe laft were alfo, in their turn, to give place to the philofophy of fome later period. This is, however, a conclufion that hardly any one will now be bold enough to maintain, after a hundred years of the moft fcrupulous examination have done nothing but add to the evidence of the NEWTONIAN SYSTEM. It feems certain, therefore, that the rife and fall of theories in times paft, does not argue, that the fame will happen in the time that is to come.

448. The multifarious and extremely diverfified object of geological refearches, does, no doubt, render the firft fteps difficult, and may very well

well account for the inſtability hitherto obſerved in ſuch theories ; but the very ſame thing gives reaſon for expecting a very high degree of certainty to be ultimately attained in theſe inquiries.

Where the phenomena are few and ſimple, there may be ſeveral different theories that will explain them in a manner equally ſatisfactory ; and in ſuch caſes, the true and the falſe hypotheſes are not eaſily diſtinguiſhed from one another. When, on the other hand, the phenomena are greatly varied, the probability is, that among them, ſome of thoſe *inſtantiæ crucis* will be found, that exclude every hypotheſis but one, and reduce the explanation given to the higheſt degree of certainty. It was thus, when the phenomena of the heavens were but imperfectly known, and were confined to a few general and ſimple facts, that the Philolaic could claim no preference to the Ptolemaic ſyſtem : The former ſeemed a poſſible hypotheſis ; but as it performed nothing that the other did not perform, and was inconſiſtent with ſome of our moſt natural prejudices, it had but few adherents. The invention of the teleſcope, and the uſe of more accurate inſtruments, by multiplying and diverſifying the facts, eſtabliſhed its credit ; and when not only the general laws, but alſo the inequalities, and diſturbances of

 the

the planetary motions were underſtood, all phyſical hypotheſes vaniſhed, like phantoms, before the philoſophy of NEWTON. Hence the number, the variety, and even the complication of facts, contribute ultimately to ſeparate truth from falſehood; and the ſame cauſes which, in any caſe, render the firſt attempts toward a theory difficult, make the final ſucceſs of ſuch attempts juſt ſo much the more probable.

This maxim, however, though a general encouragement to the proſecution of geological inquiries, does not amount to a proof that we are yet arrived at the period when thoſe inquiries may ſafely aſſume the form of a theory. But that we are arrived at ſuch a period, appears clear from other circumſtances.

449. It cannot be denied, that a great multitude of facts, reſpecting the mineral kingdom, are now known with conſiderable preciſion; and that the many diligent and ſkilful obſervers, who have ariſen in the courſe of the laſt thirty years, have produced a great change in the ſtate of geological knowledge. It is unneceſſary to enumerate them all; FERBER, BERGMAN, DE LUC, SAUSSURE, DOLOMIEU, are thoſe on whom Dr Hutton chiefly relied; and it is on their obſervations and his own that his ſyſtem is founded. If it be ſaid, that only a ſmall part of the earth's ſurface has yet been ſurveyed, and deſcribed with

with ſuch accuracy as is found in the writers juſt named, it may be anſwered, that the earth is conſtructed with ſuch a degree of uniformity, that a tract of no very large extent may afford inſtances of all the leading facts that we can ever obſerve in the mineral kingdom. The variety of geological appearances which a traveller meets with, is not at all in proportion to the extent of country he traverſes; and if he take in a portion of land ſufficient to include primitive and ſecondary ſtrata, together with mountains, rivers, and plains, and unſtratified bodies in veins and in maſſes, though it be not a very large part of the earth's ſurface, he may find examples of all the moſt important facts in the hiſtory of foſſils. Though the labours of mineralogiſts have embraced but a ſmall part of the globe, they may therefore have comprehended a very large proportion of the phenomena which it exhibits; and hence a preſumption ariſes, that the outlines, at leaſt, of geology have now been traced with tolerable truth, and are not ſuſceptible of great variation.

450. When the phenomena of any claſs are in general ambiguous, and admit of being explained by different or even oppoſite theories; if few of thoſe excluſive facts are known, which admit but of one or a few ſolutions, then we have no right to expect much from our endeavours to generalize, except the knowledge

 of

of the points where our information is moſt deficient, and to which our obſervations ought chiefly to be directed. But that many of the excluſive and unambiguous inſtances are known, in the natural hiſtory of the globe, I think is evident from the reaſoning in the foregoing pages, where ſo many examples have occurred of appearances that give the moſt direct negative to the Neptunian ſyſtem, and exclude it from the number of poſſible hypotheſes, by which the phenomena of geology can be explained. The abundance of ſuch inſtances is an infallible ſign, that the maſs of knowledge is in that ſtate of fermentation, from which the true theory may be expected to emerge.

451. Another indication of the ſame kind, is the near approach that even the moſt oppoſite theories make, in ſome reſpects, to one another. There are ſo many points of contact between them, that they appear to approximate to an ultimate ſtate, in which, however unwillingly, they muſt at laſt coincide. That ultimate form, too, which all theſe theories have a tendency to put on, if I am not deceived, is no other than that of the Huttonian theory.

452. The firſt example I ſhall take from the ſyſtem of Sauſſure. It is to be regretted, that this excellent geologiſt has no where given us a complete account of his theory. Some of the leading

leading principles of it are, however, unfolded in the courſe of his obſervations, and enable us to form a notion of its general outline. It was evidently far removed from the ſyſtem of ſubterraneous heat, and ſeems, eſpecially in the latter part of the author's life, to have been very much accommodated to the prevailing ſyſtem of WERNER. Neverthelefs, with ſo little affinity between their general views, Sauſſure and Hutton agree in that moſt important article which regards the elevation of the ſtrata. Sauſſure plainly perceived the impoſſibility of the ſtrata being formed in the vertical ſituations which ſo many of them now occupy; and he takes great pains to demonſtrate this impoſſibility, from ſome facts that have been referred to above. He alſo believed that this elevation had been given to ſtrata that were originally level, by a force directed upwards, or by the *refoulement* of the beds, not by their falling in, as is the opinion of De Luc and ſome other of the Neptuniſts.

Now, whoever admits this principle, and reaſons on it conſiſtently, without being afraid to follow it through all its conſequences, muſt unavoidably come very cloſe to the Huttonian theory. He muſt ſee, that a power which, acting from below, produced this great effect, can never have belonged to water, unleſs rarefied into ſteam by the application of heat. But if it be once admitted that heat reſides in the mineral regions, the great objection to Dr Hutton's ſyſtem is removed; and the theoriſt, who was furniſhed with ſo active and ſo powerful an agent, would be very unſkilful in the management of his own reſources, if he did not employ it in the work of conſolidating as well as in that of raiſing up the ſtrata. A little attention will ſhew, that it is qualified for both purpoſes; though inſuperable objections muſt, no doubt, offer themſelves, where the effects of compreſſion are not underſtood. We may ſafely conclude, then, that the accurate and ingenious Oreologiſt of Geneva ought to have been a *Plutoniſt*, in order to give conſiſtency to the principles which he had adopted, and to make them coaleſce as parts of one and the ſame ſyſtem. If he embraced an oppoſite opinion, it probably was from feeling the force of thoſe objections that ariſe from our diſcovering nothing in the bowels of the earth like the remains left by combuſtion, or inflammation, at its ſurface. The ſecret by which theſe ſeeming contradictions are to be reconciled, was unknown to this mineralogiſt, and he has accordingly decided ſtrongly againſt the action of fire, even in the caſe of thoſe unſtratified ſubſtances that have the greateſt affinity to volcanic lava.

453. The

453. The theoretical conclufions of another accurate and fkilful obferver, Dolomieu, furnifh a ftill more remarkable example of a tendency to union between fyftems profeffedly hoftile to one another.

This ingenious mineralogift, obferving the interpofition of the bafalt between ftratified rocks, fo that it had not only regular beds of fandftone for its bafe, but was alfo covered with beds of the fame kind, faw plainly that thefe appearances were inconfiftent with the fuppofition of common volcanic explofions at the furface. He therefore conceived, that the volcanic eruption had happened at the bottom of the fea, (the level of which, in former ages, had been much higher than at prefent), and that the materials afterwards depofited on the lava, had been in length of time confolidated into beds of ftone. It is evident, that this notion of fubmarine volcanoes, comes very near, in many refpects, to Dr Hutton's explanation of the fame appearances. If the only thing to be accounted for were the phenomenon in queftion, it cannot be denied that Dolomieu's hypothefis would be perfectly fufficient; but Dr Hutton, to whom this phenomenon was familiar, and who, like Dolomieu, conceived the bafalt to have been in fufion, was convinced that the retreat of the fea was not a fact well attefted by geological appearances, and if admitted, was inadequate to account for the facts usually explained by it. He conceived, therefore, that such lava as the preceding had flowed not only at the bottom of the sea, but in the bowels of the earth, and having been forced up through the fissures of rocks already formed, had heaved up some of these rocks, and interposed itself between them. This agrees with the other facts in the natural history both of the basaltes and the strata.

It is plain, that, in this, there is a great approach of the two theories to one another : both maintain the igneous origin of basaltes, and its affinity to lava ; both acknowledge that this lava cannot have flowed at the surface, and that the strata which cover it have been formed at the bottom of the sea. They only differ as to the mode in which the submarine or subterraneous volcano produced its effect, and that difference arises merely from the one geologist having generalized more than the other. Dolomieu sought to connect the basalt with the lavas that proceed from volcanic explosions at the surface ; Dr Hutton sought not only to connect these two appearances with one another, but also with the other phenomena of mineralogy, particularly with the veins of basaltes, and the elevation of the strata.

454. In

454. In another point, the coincidence of Dolomieu's opinions and Dr Hutton's is ſtill more ſtriking. The former has remarked, that many of the extinguiſhed volcanoes are in granite countries, and that, neverthelefs, the lavas that they have erupted contain no granitic ſtones. There muſt be, therefore, ſays he, ſomething under the granite, and this laſt is not, at leaſt in all caſes, to be conſidered as the baſis of the mineral kingdom, or as the body on which all others reſt. In this ſyſtem, therefore, granite is not always a primordial rock, any more than in Dr Hutton's.

But Dolomieu makes a ſtill nearer advance to the Huttonian theory; for he ſuppoſes, that under the ſolid and hard cruſt of the globe, there is a ſphere of melted ſtone, from which this baſaltic lava was thrown up. The ſyſtem of ſubterraneous heat is here adopted in its utmoſt extent, and in that form which is conſidered as the moſt liable to objection, viz. the exiſtence of it at the preſent moment, in ſuch a degree as to melt rocks, and keep them in a ſtate of fuſion. In this concluſion, the two theories agree perfectly; and if they do ſo, it is only becauſe the nature of things has forced them into union, notwithſtanding the diſſimilitude of their fundamental principles.

This

This ought to be conſidered as a ſtrong proof, that the phenomena known to mineralogiſts are ſufficient to juſtify the attempts to form a theory of the earth, and are ſuch as lead to the ſame concluſions, where there was not only no previous concert, but even a very marked oppoſition. I have already obſerved, that there is a greater tendency to agree among geological theories, than among the authors of thoſe theories.

455. Another circumſtance worthy of conſideration is, that in the ſearch which the Neptuniſts have made, for facts moſt favourable to the aqueous formation of minerals, we find hardly any of a kind that was unknown to the author of the ſyſtem here explained. The appearances on which WERNER grounds his opinion with reſpect to baſaltes, and by which he would exclude the action of fire from any ſhare in the formation of it, are all comprehended in the alternation of that rock with beds, or ſtrata obviouſly of aqueous origin. Now theſe appearances were well known to Dr Hutton, and are eaſily explained by his theory, provided the effects of compreſſion are admitted. From this, and the other circumſtances juſt obſerved, I am diſpoſed to think, that the great facts on which every geological ſyſtem muſt depend, are now known, and that it is not too bold an anticipation to ſay, that a theory of the earth, which explains

explains all the phenomena with which we are at prefent acquainted, will be found to explain all thofe that remain to be difcovered.

456. The time indeed was, and we are not yet far removed from it, when one of the moft important principles involved in Dr Hutton's theory was not only unknown, but could not be difcovered. This was before the caufticity produced in limeftone by expofure to fire was underftood, and when it was not known that it arofe from the expulfion of a certain aerial fluid, which before was a component part of the ftone. It could not then be perceived, that this aerial part might be retained by preffure, even in fpite of the action of fire, and that in a region where great compreffion exifted, the abfence of caufticity was no proof that great heat had not been applied. The difcoveries of Dr Black, therefore, mark an era, before which men were not qualified to judge of the nature of the powers that had acted in the confolidation of mineral fubftances. Thofe difcoveries were, indeed, deftined to produce a memorable change in chemiftry, and in all the branches of knowledge allied to it; and have been the foundation of that brilliant progrefs, by which a collection of practical rules, and of infulated facts, has in a few years rifen to the rank of a very perfect fcience. But even before they had explained the nature of carbo-

nic

nic gas, and its affinity to calcareous earth, I am not ſure but that Dr Hutton's theory was, at leaſt, partly formed, though it muſt certainly have remained, even in his own opinion, expoſed to great difficulties. His active and penetrating genius ſoon perceived, in the experiments of his friend, the ſolution of thoſe difficulties, and formed that happy combination of principles, which has enabled him to explain the moſt enigmatical appearances in the natural hiſtory of the earth.

As we are not yet far removed from the time when our chemical knowledge was too imperfect to admit of a ſatisfactory explanation of the phenomena of mineralogy, ſo it is not unlikely that we are approaching to other diſcoveries that are to throw new light on this ſcience. It would, however, be to argue ſtrangely to ſay, that we muſt wait till thoſe diſcoveries are made before we begin any theoretical reaſonings. If this rule were followed, we ſhould not know where the imperfections of our ſcience lay, nor when the remedies were found out, ſhould we be in a condition to avail ourſelves of them. Such conduct would not be caution, but timidity, and an exceſs of prudence fatal to all philoſophical inquiry.

457. The truth, indeed, is, that in phyſical inquiries, the work of theory and obſervation muſt

go hand in hand, and ought to be carried on at the ſame time, more eſpecially if the matter is very complicated, for there the clue of theory is neceſſary to direct the obſerver. Though a man may begin to obſerve without any hypotheſis, he cannot continue long without ſeeing ſome general concluſion ariſe ; and to this naſcent theory it is his buſineſs to attend, becauſe, by ſeeking either to verify or to diſprove it, he is led to new experiments, or new obſervations. He is led alſo to the very experiments and obſervations that are of the greateſt importance, namely, to thoſe *inſtantiæ crucis*, which are the *criteria* that naturally preſent themſelves for the trial of every hypotheſis. He is conducted to the places where the tranſitions of nature are moſt perceptible, and where the abſence of former, or the preſence of new circumſtances, excludes the action of imaginary cauſes. By this correction of his firſt opinion, a new approximation is made to the truth ; and by the repetition of the ſame proceſs, certainty is finally obtained. Thus theory and obſervation mutually aſſiſt one another ; and the ſpirit of ſyſtem, againſt which there are ſo many and ſuch juſt complaints, appears, nevertheleſs, as the animating principle of inductive inveſtigation. The buſineſs of ſound philoſophy is not to extinguiſh this ſpirit, but to reſtrain and direct its efforts.

458. It

458. It is therefore hurtful to the progreſs of phyſical ſcience to repreſent obſervation and theory as ſtanding oppoſed to one another. Bergman has ſaid, "Obſervationes veras quàm ingenioſiſſimas fictiones ſequi præſtat; naturæ myſteria potius indagare quàm divinare."

If it is meant by this merely to ſay, that it is better to have facts without theory, than theory without facts, and that it is wiſer to inquire into the ſecrets of nature, than to gueſs at them, the truth of the maxim will hardly be controverted. But if we are to underſtand by it, as ſome may perhaps have done, that all theory is mere fiction, and that the only alternative a philoſopher has, is to devote himſelf to the ſtudy of facts unconnected by theory, or of theory unſupported by facts, the maxim is as far from the truth, as I am convinced it is from the real ſenſe of Bergman. Such an oppoſition between the buſineſs of the theoriſt and the obſerver, can only occur when the ſpeculations of the former are vague and indiſtinct, and cannot be ſo *embodied* as to become viſible to the latter. But the philoſopher who has aſcended to his theory by a regular generalization of facts, and who deſcends from it again by drawing ſuch palpable concluſions as may be compared with experience, furniſhes the infallible means of diſtinguiſhing between *perfect ſcience* and *ingenious fiction.* Of a geological theory that

that has ſtood this double teſt of the analytic and ſynthetic methods, Dr Hutton has furniſhed us with an excellent inſtance, in his explanation of granite. The appearances which he obſerved in that ſtone led him to conclude, that it had been melted, and injected while fluid, among the ſtratified rocks already formed. He then conſidered, that if this is true, veins of granite muſt often run from the larger maſſes of that ſtone, and penetrate the ſtrata in various directions; and this muſt be viſible at thoſe places where theſe different kinds of rock come into contact with one another. This led him to ſearch in Arran and Glen-tilt for the phenomena in queſtion; the reſult, as we have ſeen, afforded to his theory the fulleſt confirmation, and to himſelf the high ſatisfaction which muſt ever accompany the ſucceſs of candid and judicious inquiry.

459. It cannot, however, be denied, that the impartiality of an obſerver may often be affected by ſyſtem; but this is a misfortune againſt which the want of theory is not always a complete ſecurity. The partialities in favour of opinions are not more dangerous than the prejudices againſt them; for ſuch is the ſpirit of ſyſtem, and ſo naturally do all men's notions tend to reduce themſelves into ſome regular form, that the very belief that there can be no theory, becomes a theory itſelf, and may have no inconſiderable ſway

ſway over the mind of an obſerver. Beſides, one man may have as much delight in pulling down, as another has in building up, and may chooſe to diſplay his dexterity in the one occupation as well as in the other. The want of theory, then, does not ſecure the candour of an obſerver, and it may very much diminiſh his ſkill. The diſcipline that ſeems beſt calculated to promote both, is a thorough knowledge of the methods of inductive inveſtigation; an acquaintance with the hiſtory of phyſical diſcovery; and the careful ſtudy of thoſe ſciences in which the rules of philoſophizing have been moſt ſucceſsfully applied.

FINIS.

Printed in the United States
By Bookmasters